the
trees
are
speaking

the
trees
are
speaking

DISPATCHES FROM
THE SALMON FORESTS

lynda v.
mapes

University of Washington Press / Seattle

Design by Mindy Basinger Hill / Composed in 11.8/16 pt Bely Regular

29 28 27 26 25   5 4 3 2 1

Printed and bound in the United States of America

Photographs by the author unless otherwise noted.
Page ii: KLBahr / istockphoto.com
Page viii: Petra Richli / AdobeStock
Page 204: David Herasimtschuk.

UNIVERSITY OF WASHINGTON PRESS *uwapress.uw.edu*

LIBRARY OF CONGRESS CATALOGING-IN-PUBLICATION DATA
Names: Mapes, Lynda, 1959– author.
Title: The trees are speaking : dispatches from the salmon forests / Lynda V. Mapes.
Description: Seattle : University of Washington Press, [2025] | Includes
    bibliographical references and index.
Identifiers: LCCN 2024053256 | ISBN 9780295753676 (hardcover) |
    ISBN 9780295753683 (ebook)
Subjects: LCSH: Old growth forest ecology—North America. | Old growth forest
    conservation—North America. | Salmon—Habitat—North America. |
    Nature—Effect of human beings on.
Classification: LCC QH102 .M37 2025 | DDC 577.3097—dc23/eng/20250122
LC record available at https://lccn.loc.gov/2024053256

∞ This paper meets the requirements of ANSI/NISO z39.48-1992
    (Permanence of Paper).

FOR THE TREES

CONTENTS

OURS IS A TIME OF RECKONING for our forests. What is being wrought in our forests everywhere, now, even in parks and protected old-growth reserves, is an apocalypse of hot drought, fire, bugs, pestilence, and death. What we humans have unleashed is beyond anything in our experience—and so far, our ability to manage. We need to rethink our relationship with our forests. Nothing less than a paradigm shift will save them. This is necessary, even lifesaving work. We will survive only if our forests do, with their cooling, life-giving canopies, their water-purifying roots, and their air-cleansing leaves. About a third of the carbon dioxide gas cooking the planet that we spew from burning fossil fuels is transubstantiated by terrestrial plants, including trees, by the miracle of photosynthesis, never replicated in a scientist's lab. In an alchemy of sunlight, water, and air, trees transform the carbon in carbon dioxide and water into the food that grows their wood, roots, and leaves. The by-product of their work is deep soft shade, as trees grow and lift and interlace their canopies; pure oxygen for us all, providing the ark of refuge and gifts of abundance that are our forests.

The health benefits of trees to humans, their essential link to the survival of species, their moderation of our climate catastrophe, are so essential that if forests did not exist, arguably neither could we—at least not in ways we can imagine or ever want. This is why we must protect what forests remain and learn how to use them differently. Or more accurately, for most of us, *relearn* how to use them. This means taking on the hard fact of our relentless and heedless consumption. It means taking to heart the teachings of Western science about how forests work and what they need. It means listening to and learning from the teachings of the First Peoples who live in a relationship of reciprocity and interdependence with forests.

This book began for me as an encounter with the ecology of old-growth forests in the Pacific Northwest—the science of what they are, why they matter, how they work. I explored some of the oldest trees alive, on Vancouver Island in British Columbia, part of the largest temperate rain forest on Earth, and especially the *salmon forests*. The term is used by author and environmental activist David Suzuki, among others, to describe the swath of temperate rain forest nurtured by the largest animal migration in the North Pacific: salmon coming home from the ocean to their freshwater streambeds, where they were born. Here is an interconnected abundance so intimately interdependent—the forests laced with freshwater streams are full of salmon, and the salmon are nurtured by the forests. The beautiful term *salmon forest* is not a mystical concept but a scientific truth, documented by decades of research. The influence of salmon on these forests is so profound, John Reynolds of Simon Fraser University learned, that the signal of the nutrition they bring back from the sea can be seen in a surge of canopy greenness visible from space. Reynolds worked with his collaborators to examine streamside vegetation in fifty watersheds in the Great Bear Rainforest of British Columbia and found the plant community along salmon streams is dominated by species such as salmonberry that indicate rich soils. Salmon, this research showed, influence not only the abundance but the community structure of the streams and forests where they are born and where they return and die. In death, they are transformed into the flesh of bears, the succulence of salmonberries, and the enormity of trees surging with nutrition from salmon.

Tom Reimchen of the University of Victoria, British Columbia, pioneered the science of tracking marine-derived nutrients from salmon, by identifying their unique isotopic signature. He found marine nutrients in birds and bugs feeding on spawned-out fish hauled into British

Columbia's old-growth forests by wolves and bears. He found fatter rings in those trees too, indicating more growth during the years of big salmon returns. This is an intimate connection between land and sea, a living abundance mediated by wildlife fed by salmon. They are the manna of the sea, wealth not kept in a bank but in the land and the streams nurturing the next generations of salmon, people, and wildlife as well as the forests that shelter them.

These forests are found amid the old-growth temperate rain forests on the western slopes of the mountains from Southeast Alaska to British Columbia through Washington, Oregon, and Northern California. Steeped in as much as twelve to fourteen feet of rainfall, these forests thrive in moderate temperatures in low-elevation valleys. This is the land of the light sweater, the layers peeled but never put too far away; it doesn't freeze here often, nor do summer temperatures often rise above 80 degrees. Apart from the wetness and their mild climate, it is the sheer botanical beneficence of these old-growth temperate forests that sets them apart. These are forests with trees so massive and so rich with vegetation, they are among the most carbon-dense forests on Earth. The trees are plastered with even more green life growing atop them—mosses and lichens drape and festoon and coat tree trunks and branches, and fern gardens sprout and cascade from treetop to ground.

To be in an old-growth forest is to feel cloaked, as if walking in a living terrarium, padding around a soft kingdom of green. Sitka spruce, western hemlock, and cedar interspersed with bigleaf maples garlanded with ferns and swags of moss are the signatures of these forests. Some of the conifers can persist to great age: Douglas fir to five hundred, eight hundred, and even a thousand years, and cedar even longer. These trees can tower as much as 250 feet in height and grow to 60 feet around. These are forests with dead and downed logs everywhere. Communities of mosses, tree seedlings, and fungi thrive in this dead wood, which is more alive than the living trees next to it, for sheer biomass. Small mammals, amphibians, birds, and bears all make homes in the cavities of these dead and decaying trees. For old-growth forests are places where everything is alive, even things that supposedly are not—from the rocks cushioned with lichen and moss to the snags and logs teeming with new life.

Jerry Franklin, the eminent forest ecologist and national authority on sustainable forestry practices, with his colleagues, helped reveal much of what we know today about old-growth forests. Over the course of Franklin's lifetime—he is something of an old-growth tree himself—his work has been the stuff of breakthroughs, to understand that there *is* such a thing as an old-growth forest. That this very particular type of forest is without equal, not just a bigger version of a younger forest but unique in its attributes that are only gained over long cadences of time. Here is a diversity of structure and complexity and a completeness that takes centuries to form. Not only the gigantic trees but their shattered and broken tops, their cavities that shelter so much life, their streamers of bark that birds and bugs tuck under, their broad branches for forest birds that make no nest, such as the marbled murrelet, which flies all the way to the sea to provision its young. The flying squirrels and tree voles that thrive in the canopy and the spotted owls that used to—until we cut down so much of these old-growth forests and invited, with our many alterations of the landscape, the invasive barred owl. Here too is so much dead wood, lounging on the forest floor and standing as snags inviting roosting, nesting, and perching animals, and nurse logs, sprouting in their juicy rot a thick fur of tree seedlings. Here is a marvelous diversity of landscape: old-growth forests are not just comprised of big trees but open areas too, where an elder has crashed over, clearing the way for the next generation. There is a lot of food there, in the berries and young surging forest and thriving understory plants.

During many walks and talks and long reflections over the years with Franklin, I heard how he and his collaborators learned all this. There was much more to it than data collection or experiments; this was about listening to the trees, with a stance of humility, willing to be instructed by the forest itself. Over his long career Franklin loved to convene collaborators across disciplines in what he called "pulses" to camp together for days or weeks and ask profound questions about everything—much of which had not been explored. From the immense amount of dead and downed wood they saw on the forest floor, to the logs tumbled into streams in old-growth forests, to the mosaic of age classes and variety of landscapes in these forests. This was a patient learning and listening for what the old trees could teach about how these forests are structured, how they work, and how they renew themselves.

There is some irony in this: Franklin initially thought he wouldn't study Douglas fir because everyone already knew all about it. But he quickly found out we didn't know anything at all about these moist old-growth Douglas fir and western hemlock forests. At the US Forest Service in the post–World War II era and for decades after, these old-growth stands were regarded and treated as biological deserts, decadent wastelands, the sooner gotten rid of the better, to make way for thrifty young industrial plantations. Fiber farms, clear-cut, sprayed with herbicide, then planted in a monoculture for the next cut in forty years or even less. The revolution in this thinking—partly as a result of the scientific understanding Franklin helped ignite and promote in public policy—cannot be overstated, culminating in the protection of millions of acres of old-growth forests in 1994 on federal lands in Washington, Oregon, and Northern California. Yet today, especially on industrial lands, the clear-cut is still alive and well. Franklin showed me some of those clear-cuts, and when he did, he was angry. For these were not *forests* but *farms*. And as they suck water from and displace natural forests and old trees, these fiber farms are fueling an apocalypse of bugs, hot drought, and fire. They represent primarily one value: efficiently making as much money as possible from trees grown as a commodity crop. What streams from these trees is not lichen and ferns—they never grow old enough for that—but profit, sent to distant investors managing real estate investment trusts (REITs), insurance companies, hedge funds, university endowments, even our own pension plans.

One day in particular that I spent with Franklin set the direction for this book. We had shared a morning exploring the Cedar Flats Research Natural Area, a spectacular old-growth forest in the southwestern Cascades, a place Franklin had taken students for decades, so textbook perfect is this forest's splendor and old-growth attributes. Here was everything from the wizened mother cedars silvered with age and splintered tops, spearing the sky, to the Douglas fir with broad branches perfect for a murrelet nest and a trunk wide as a garage door. The delicate forest orchids, lush ferns, and sun shafting through clearings in the canopy and mossy nurse logs. We enjoyed all this together, and then Franklin suggested we take a ride down the road. He wanted to show me something.

Just a short distance from the magnificence of Cedar Flats was a sharp-edged clear-cut, set on the landscape with a surveyor's precision. The rising green of the old-growth we had just explored could be seen in the distance. But here, where we stood among shattered stumps and searing sun where a few planted trees struggled to live, was an industrial clear-cut, just like so much had already been cut into this landscape.

In a world fraught with climate warming and species on the verge of extinction, Franklin said, this was the kind of thing we should not be doing any more. Places like this—doing nothing for biodiversity, for carbon sequestration, for water management, for growing the old-growth of tomorrow. No more farms like these should be made from natural forests anywhere, he said, and certainly no more old-growth anywhere should be lost. I was startled. The impetus for this book project, what I thought I was doing, shifted.

Under the hot sun in that cut-up landscape, my work of exploring old-growth forests, and their beautiful dynamics, also became a hard learning of the history of how we've lost so much of them. In addition to blowing up the topic, I exploded the geography. I had intended to journey the Pacific Northwest's drip line of old-growth forests and be done. That didn't happen. I could not help it, after the disruptive and perhaps inevitable experience of confronting the old-growth forests of Oregon, Washington, and British Columbia—and seeing how little remains. On the United States side of the US-Canada border was a mother lode of big trees, fully a quarter of the old-growth left in this country. On *both* sides of that border, alongside the soaring grandeur, were tatters. Amid Native knowledge systems of sustained intergenerational management, where did the wholly alien idea come from to cut down—to *annihilate*—the living wealth of these forests and send it somewhere else for the mere ephemera of money?

The book therefore became a witness not only to the incredible ecological and cultural values of these forests but to the connected history of their loss, beginning on the East Coast in Maine and repeated across the United States and continuing over the border into Canada, even today. Along both the Pacific and Atlantic coasts this was a ransacking of forests by people who had already cut down the forests of Europe. They arrived first along the northeastern coast of what is now the United States, timber-famished on verdant shores, claimed Native lands as their own, and set to them with saw and plow. By the time Captain James Cook showed up in 1778 at Friendly Cove on Nootka Sound, in what would become British Columbia, the logging of Maine's old-growth white pine forests had been underway since the first cutting on Monhegan Island in the 1600s. Bangor, Maine, was well on its way to becoming the world's largest lumber port. Much of New England had been cleared for agriculture or cut over before the American Civil War, before Washington was even a state.

These were the Atlantic salmon forests, nurturing eleven spectacular, interdependent runs of sea-run fish and generations of the Wabanaki, ancestors of the Abenaki, Maliseet, Micmac, Passamaquoddy, and Penobscot Nations. Historians believe these First Peoples have lived in the area of what is now northeastern New England and maritime Canada for more than twelve thousand years. While never as numerous as the salmon in the Pacific Northwest, these fat, glorious fish were part of the abundance of the rivers that nurtured alewife by the millions, shad, blueback herring, sea lamprey, sturgeon, and more—a banquet of food present through nearly all seasons. The cut-and-run ethic that leveled these forests and nearly exterminated the salmon and their silvery cohort was replicated in at least one instance on both the Atlantic and the Pacific coasts by the *same family*. They stripped the best timber around Machias, Maine, then set up one of the longest continuously operating sawmills in North America, shut down in 1995, in Port Gamble, Washington, on Puget Sound.

On foot, by boat, by seaplane, by logging road and bushwhack, in the salmon forests of both coasts, I sought this story. I explored miles of forestlands, some cut to bits, some still intact, whole and wondrously alive, from the crowns of five-hundred-year-old trees to the depths of their interconnected roots. I talked with scientists, Native Americans, First Peoples, historians, loggers, and activists. I climbed a four-hundred-year-old Douglas fir, probed the sagas of salmon, Pacific and Atlantic, and rivers both dammed and finally running free. The only spotted owls I saw were stuffed. I witnessed forests in Oregon aflame that had not burned in five hundred years. I grieved just-logged old-growth trees on Vancouver Island, British Columbia, their stumps still wet with sap and wide enough to lie across.

I met Native leaders in a small but mighty band determined to reassert their right and title to their lands as well as their simple but powerful vision of governance: *Hisuk-ish tsa'walk* (everything is connected). I met archaeologists revealing intense management of old-growth forests through generations of First Peoples, curating cedars used for everything from house planks to clothing to canoes. Nothing was wasted—the trees were partially stripped of bark and even planked in such a way as to leave the tree standing, and alive, for the next harvest. From forest gardens of crab apples and intertidal root gardens and stone salmon traps, I learned about Indigenous management systems designed to leave the biggest fish, plenty of salmon to spawn, and use the same forests in their home territory to provide everything people needed (for food, clothing, and shelter) *for generations.* The exact opposite, in other words, of cut and run.

Along Maine's Penobscot River, I met scientists and tribal leaders working to bring sea-run fish, including Atlantic salmon, back to the state's largest river. I saw river herring, gone from their home streams for more than a century, surging back to their spawning lakes. I saw the river, running free and full where dams had been taken down in a collaborative effort to restore the natural wealth of the Penobscot, for so long harnessed mainly for power and industry. I saw Atlantic salmon swagger into fish windows at a dam, the biggest thing around, pure silver muscle whose scientific name is *Salmo salar*, the leaper, popularly known as the King of Fish, yet still no match for this dam. The Penobscot is the last stronghold in North America of Atlantic salmon. Among the most endangered fish in the country, they were nearly lost to dams, pollution so profound the river turned colors, and mauled spawning tributaries, turned into chutes for log drives.

I saw East Coast old-growth forests in late stages of decline even in protected areas, with big trees dying before their time and entire species lost, or on their way out, victims of bugs and fire that know no political boundaries. This is the predictable result of the intertwined catastrophes of global commerce and the carbon bomb humans continue to detonate every hour, every day, around the world, into our shared atmosphere. To log or not to log at least is under our control. A choice. But the forces of death and destruction in today's forests, even *in reserved, protected areas,* will not be quelled without doing much, much more than deciding where, how much, and how we will log.

Exploring all of this, I witnessed much that saddened me. But I also met brave people of every sort, deeply committed to healing their home. It was there, with them, even amid the unsolved problems and gnawing questions, that I felt hope. Neither they nor I can offer a sunbeam-gilded path to fixing it all. But I found people unafraid to create ways to persist in their chosen home. Their stories raised uncomfortable questions about this, our cutover land, in the four hundredth year of our unplanned experiment with the commodification of everything—even one another. The glorious salmon on the Atlantic and Pacific coasts and all their cohorts—herring, lamprey, and all the rest—are gifts as gorgeous and essential as the forests that shelter and nurture them. Despite everything we have done to them, they are still here. So are Native Americans and First Nations people, many with knowledge of how to live among these landscapes that others among us didn't even know to miss.

The need for a paradigm shift is readily apparent. The trees are speaking, and telling us, wrote coauthors Michael Paul Nelson and Tom DeLuca, "the climate we have created and the forests we have created cannot co-exist. . . . The forests are reminding us that the fundamental reality of the world is change, not stasis." After a 2023 wildfire burned the H. J. Andrews Experimental Forest outside Eugene, Oregon, much of which had not seen fire in at least five centuries, they wrote: "Even longstanding symbols of resiliency—towering Douglas-fir trees, some 800-years-old—have fallen to the reality of climate change. Simply cordoning off some parts of the world from Western voracity will not solve our crisis, protect those places or put us in right relationship with the world. The forest is asking us to find ways to respectfully see the forest and the human as part of a larger shared and interdependent community, engaged in an active and reciprocal relationship to enhance their resilience, and in turn our own."

In this moment there is much work to do and even, to be sure, much to dread. But there's also this, from which to take inspiration: the immense capacity for renewal within the lands and waters themselves. For here is the thing: If enough of their biological legacies are protected, and natural processes are allowed to revive, rivers and forests and the life they nurture can renew themselves. Emerge, along with us humans, to a new start. I saw it. I know this capacity for renewal to be true. As we confront biological impoverishment and climate catastrophe, this restorative power is what we must unleash. We must fix what we can, make space for Nature to renew, and stop wrecking more.

But we are getting ahead of our story. Let us take a step back, into a lush vision of perfection. The old-growth forests of the Pacific Northwest. Let's start there, in the moss and ferns, the lichen, and take the counsel of cool, wet, colossal old trees. The rest, well, it will unfold. By the time we are done, the recognition and respect due from our adolescent species for our shared fate with all creation will be clear. That, and something the First Peoples have known all along: This is our only home.

the
trees
are
speaking

1

COMBING THE HIGH WINDS

TREES SOAR INTO THE CANOPY out of sight, more than two hundred feet aboveground. Even at some four hundred years old, these forest giants have not yet reached midlife. A Pacific wren's arpeggios exude from the boughs of a vine maple, and the silvery wisp trailed by a ballooning spider glints in the light beaming through a wide gap in the canopy. The buzzy call of a thrush comes from somewhere deep within many shades of green.

This forest is a rare 680-acre island of old-growth amid a sea of plantation clear-cuts in the Gifford Pinchot National Forest, in southwestern Washington State. Set aside as a forest reserve in 1946 by the US Forest Service, the Cedar Flats Research Natural Area is perched on a gently undulating series of bluffs following the curves of the Muddy River, so named for all the sediment and glacial flour rinsing down from the slopes of Mount St. Helens, the most active volcano in the Cascade Range. Here the climate is wet and cool, with the rains measured in feet, peaking in the winter and tapering to very low rainfall in the summer, typical of the Pacific Northwest climate. People not from here say it rains all the time, but the summer brown lawns of the towns speak a different truth.

This reserve is one of the finest examples of an old-growth forest left anywhere on the western slope of the Cascades. Here abide the grand species that are the signature of these forests: Douglas fir and western redcedar, capable of living longer than one thousand years and persisting as dead and downed logs for centuries as they gradually decompose into the soil. These species—capable of growing to immense height and size in the mild and moist Cascade forests from

British Columbia to Northern California—infrequently see fire. Cedars more than 10 feet around at breast height (far larger at the base) and more than 250 feet high keep company with their companions of red alder, vine maple, and bigleaf maple. The landscape is diverse: a swampy marsh at the edge of the reserve boasts giant redcedars with gray broken tops spearing the low clouds. In the swampy ground there are lush sedges, small fruited bulrushes, and black cottonwoods that sweeten the spring air with the floral perfume of their sap, sticky at the base of juicy new buds. Skunk cabbage lights the spring days with its bright yellow spathe that earns its common name of swamp lantern.

The forest is alive with scents: the sharp cleansing bite of cedar and fir, the deep plummy perfume of wet dead wood, and the hummus fragrance of soil so rich and chocolaty it looks nearly edible. It is a place not dense and dreary but ashine with sunlight beaming through broad gaps in the canopy, natural gaps created by deadfall and windthrow in the forest. This complexity is the mark of forests allowed to persist without human disturbance through long cadences of deep time, the trees left to live out their slow pageantry of succession. In addition to the most abundant species of Douglas fir, western redcedar, and hemlock, other conifers are resplendent: Pacific yew, Pacific silver fir, grand fir, western white pine.

Large areas of the forest are not trees at all but a mosaic of shrubs, herbs, and flowers. Glades of wild ginger, bunchberry dogwood, and club moss grow low to the ground. Bellflower and twinflower find the sun patches along with red elderberry, spirea, salmonberry, thimbleberry, and devil's club. The deepest shade has its adornments, from the velvety carpets of oxalis to Indian pipe, ghostly and white, rising in the densest shade, taking its nutrition from roots belowground. The lacy leaves and nodding purple flowers of Pacific bleeding heart, and the pristine white of trillium are grace notes of spring. Delicate white goblets of queen's cup lily, with their paired glossy leaves, light the forest floor from May through midsummer. All year long a queen's garden of ferns graces the understory: maidenhair fern, sword fern,

Pacific Northwest

lady fern, licorice fern, oak fern, deer fern. Ferns everywhere, even growing atop nurturing stumps. Some ferns will stand nearly shoulder high by the end of their growing season, as they rise in a high tide of green from the forest floor.

Salal also is everywhere abundant, with its green leathery leaves and waxy pinkish white bells that festoon its spring branches, ripening to fat purple berries by fall. The sharply-pointed glossy leaves of Oregon grape are interspersed with clusters of bright, sweet yellow flowers in spring and laden with fruit in fall. There is food everywhere in this forest—for pollinators, birds, and insects. Black bear, deer, Roosevelt elk, coyote, and bobcat. Mink, river otter, and beaver have their run of the Muddy River, and the quiet, still waters of the swamp. The banquet for small mammals, insects, and grubs never stops amid the dead and downed logs and the thickety-thick of the understory plants. Standing dead snags of the trees keep the buffet coming. There is something to eat in every season—from the starchy roots of skunk cabbage in earliest spring, and sweet juicy inner bark of the trees, to berries from as early as June through fall. Huckleberries, tiny and sweet; thimbleberries, fragile and tart; and salmonberries that draw the ascendant song of Swainson's thrush to these woods in high summer.

The old trees provide unique habitat; they are caved and drilled with cavities, home to many lives throughout the seasons, whether for a season's hibernation, a quick snack, an afternoon snooze, or a perch for snagging a meal. These old trees are an invertebrate paradise too. Banana slugs long as a finger lounge in the wet. Tree snails stud the trees, and salamanders tuck into the duff. Spring brings the call of tree frogs, shrill and insistent from cool, shaded wetlands. A calypso orchid lights the gloaming beneath a gnarled cedar, its magenta and white blossom a vision of tiny perfection, from its striped throat to the delicate pink spray of its bloom. This orchid's common name is fairy slipper. If there are fairies here, I am pretty sure this is where they live, donning fairy slipper blossoms for midnight ballets amid queen's cup lilies and lady ferns on full-moon nights.

A giant fir has fallen, making way for the next generation, in a continuity of life in this old-growth forest that links its past to its future, age upon age. Even in my booted feet, I sense the tread underfoot, its deep softness, the duff of the forest piled thick and smoothed by time. Yet off-trail there is no easy walking across this forest. The whales of massive decomposing logs are everywhere, diving ever so slowly into the ground. They invite riding but for their softness—I've learned by now these ancients will crumble at my weight. Just to test and feel what the passage of so much terrestrial time feels like, I sink a finger into a log's moist, thick inner sanctum, where this Douglas fir is living its second life as a downed log. *Living, dead,* these words don't really make sense in an old-growth forest of the Pacific Northwest, where a live tree is mostly dead but for the growing live layer of cells under its bark called cambium. And what does *dead* mean for a decomposing log that holds even more living beings than this tree did when it was alive?

My finger sinks to the knuckle in moss that covers its venerable bark, and a colonnade of young trees parades along its swaddling thickness. Some of these baby trees are no bigger than seedlings; one shows its very first whorl of leaves: a hemlock getting its start. Aptly named nurse logs like this provide a boost for the next generation, raising them up above the competition for space in the understory, where every bit of the funk of the forest floor is teeming with life. Herbs, club mosses, trailing blackberry.

I've always been drawn to these ancient forests. It is their timelessness, that continuity of a place here so long before us, abiding until long after we are gone that draws me, the sense of a touchstone of quiet, of wonder and awe. So I was thrilled when not long after that ramble in the old-growth at Cedar Flats, I got an opportunity to actually climb a massive Douglas fir, in a great sweep of old-growth. That was thanks to Mark Schulze, director at the H. J. Andrews Experimental Forest, east of Eugene, Oregon. Known simply as the Andrews to all who love it, this is a special place for learning, scientific research, arts, and the humanities. I'd been coming here for years, after scientist Fred Swanson, retired from his work as a research geologist with the US Forest Service Pacific Northwest Research Station in Corvallis, invited me to see what climbing a tree at the Andrews would be like. He knew I had done some climbing at the Harvard Forest in Petersham, Massachusetts—four thousand acres of woodlands, fields, trails, old stone walls, streams, and ponds. Both of these research forests are part of the Long Term Ecological Research Network (LTER). These sites, funded by the National Science Foundation, include twenty-eight locations over a wide variety of ecosystems—from Arctic regions to prairie to coastal Everglades to the urban ecosystem of Baltimore, Maryland, to temperate lakes and eastern woodlands. They are more than just research sites, they are communities dedicated to long-term place-based research that crosses disciplines to bring different minds and specialties together for scientific, philosophical, artistic, and cultural inquiry.

At the Andrews, I felt fortunate to be welcomed as a writer and researcher, to learn from all that these trees, and the people working here, could teach. The Andrews was established in 1948 by the US Forest Service and in 1980 became a charter member of the LTER. It encompasses part of the green forest cloak rumpled over the Earth west of the Cascade mountains from Vancouver, British Columbia, to Arcata, California, nurtured by a mild climate with wet winters. In these moist forests, clouds cruising in from the Pacific snag and pile and drop their moisture on the west side of the Cascades, drenching the forest all winter in fog, rain, and snow. Spring is long, cool, and wet. So is fall. It is, in other words, the perfect place for growing trees to truly spectacular size. The tallest known Douglas fir at the Andrews is 306 feet tall—taller than the Statue of Liberty measured from the base. The cedars too are monumental, cruising to 200 feet and fattened to 20 feet in diameter and living even a thousand years.

It is the lushness that overwhelms—the growth of green on every surface, as if a hiker pausing too long would start to green up with moss. The light even on bright summer days is deeply shaded, with sun shafts piercing gaps in the canopy. The ground is padded by the needlecast of centuries. This softness everywhere invites napping, propped on the plush give of a rotting stump. Yet there is always work going on here. The hallmark of this research forest is long-term experiments underway for decades, to learn how the forests and streams work and why they matter. Artists, photographers, writers, scientists, poets, philosophers, and students all work and visit here, one of the most studied old-growth forests on Earth. That doesn't mean the Andrews hasn't been cut, however; about 25 percent of the forest has been logged. Clear-cuts are just some of the manipulations done by scientists at this research forest in their quest to understand it.

I had arrived at the Andrews the evening before our climb, just as the first coolness was settling on the land with its fragrances of earth, streams, and trees. The wind was calm, and it was a perfect August night, the sky pellucid and saw-toothed at its horizon with the tops of old-growth firs. After enduring six hours of highways from my home in Seattle, my first impression of the forest was the quiet, so deep it was dimensional. There was not even the sound of wind. The scent of fir trees and conifer sap was everywhere, seeping from nearly sixteen thousand acres of forest. As I walked the Discovery Trail at the Andrews and heard the soft sounds of Lookout Creek, one of the clear, cold streams threading this forest, dragonflies hawked for bugs and the day's last sun painted the tops of the tallest trees gold. A distant raven *gronked*. I felt myself quieting, in this quiet.

Foamflower, graceful and demure, fizzed white and green along the edges of the trail. I saw metal bands tensioned around trees, spring-loaded to sense and record the diurnal tide of the forest. Especially in the dry season, trees reach their maximum circumference in early morning, after pulling up water overnight in their trunk and before the sun ignites another round of photosynthesis and water loss from the canopy. In this green, soft, and padded realm, there was no human noise to intrude on these sounds. I found myself working to extend my hearing as distantly as I could from where I stood, to savor the forest sounds. In a world in which I so often wince and plug my ears and rush past noise, I was aware of how different—and delicious—it was to slow, to be as still as I could, to linger, giving all my attention to listening. I felt a soft swell of something I had not felt in some time: wonder. Two tiny fawns, their supple bodies flecked with white spots, startled me from my reverie as they bounded across the trail in front of me, white tails flashing.

The Andrews forest reigns supreme over the entire drainage basin of Lookout Creek in the western Cascade Range of Oregon. Most of the forest is old-growth, with trees four hundred years and older cloaking 40 percent of the land. The old-growth conifers here— cedar, hemlock, and ancient Douglas fir, catapult any visitor into a different world and relationship with time. There are truly ancient trees here, persisting for three hundred, five hundred, and a few for even seven hundred years. The average annual rainfall here of seven feet sops and seeps into the ground. Not for nothing is the mascot of the Andrews the rough-skinned newt. I noticed instruments off in the forest, silently doing their work: probes, data loggers, gauging stations, recording everything from temperature to streamflows, de- composition, microclimate, soil moisture, and much more. In a world of short-term, instant everything, the Andrews is all about the long view, investigating the movements of animals, birds, and insects as well as the length of salamanders and the growth of fish, the daily wa- ter balance in individual trees, the distribution of light in the canopy, and even the activities of microscopic fungi in the spaces between the cells of leaf surfaces.

A living laboratory, scientists at the Andrews have recorded some of the longest data sets anywhere, with seventy-five years of research across multiple disciplines, Matt Betts, the lead principal investigator for the LTER program at the Andrews, told me. A professor of forest ecosystems and society at Oregon State University in Corvallis, Betts explained that most ecological research lasts only two or three years,

Old-growth Douglas fir has deeply furrowed bark, such as on this tree at the H. J. Andrews Experimental Forest, which could be as much as seven hundred years old. The fan-shaped branching is also a characteristic of great age.

but at the Andrews, scientists have studied tree growth and death in the same stands for fifty-two years; examined fish populations in the same section of stream for thirty-seven years; and measured climate and streamflow across the forest for sixty-five years. That's a long baseline against which to measure and understand environmental change.

THE MORNING OF MY TREE CLIMB with Mark Schulze broke gray and muggy. In my apartment at headquarters I made my standard monster field breakfast: a hunk of butter sizzled in a skillet to fry two eggs flipped onto buttered toast while still hot enough to melt a slice of cheese, all of it marvelously mounded with mayonnaise. This I washed down with coffee fierce and hot. I was ready. I heard Schulze walking in his apartment overhead, so I knew he was up and getting ready for our 8 a.m. rendezvous. A few minutes later, I opened my door and there he was, cheery and ready to go, carrying two big bags brimming with climbing gear. We hauled them out and heaved them

into the back of Schulze's pickup. It's a short drive on a woods road
from headquarters to the trail that leads to the Discovery Tree—the
tree we would climb, so named because it is used by scientists and
visitors at the Andrews forest for every sort of inquiry.

Schulze and I hiked down a short trail from the road, hauling our
bags, and suddenly there it stood. It was big all right, bigger than
I'd remembered from previous visits. The Discovery Tree hummed
with equipment arrayed all around it, cables everywhere on the forest
floor. Schulze had rigged the tree for climbing the night before, and
the bright colors of our ropes spilled down the trunk, our pathway
to the top, wherever it was, somewhere up there out of sight. I tried
to look nonchalant.

The climb began in a flurry and clatter of gear: clamping and snapping, straps and buckles. I have always hated this part of tree climbing,
the sudden alien rush of equipment a reminder of the unnaturalness
of this act. I'd climbed trees with rope and harness for research before—but those trees were less than half the size of the behemoth I
was about to meet. Soon the rhythm of equipping for a climb came
back: helmet, harness, and climbing rope, threaded through an ascender strapped to my foot. Things humans need to visit the realm of
a big tree. There are many methods for climbing, using an ascender as
I chose to do was a good choice for so high a climb. There was more
here to take on than a simple fling of a loop of climbing rope around
the tree and walking up its trunk. The trees I'd climbed before, East
Coast oaks and sugar maples at the Harvard Forest, seemed cream
puffs compared to this. They were all velvety leaves and compara-
tively smooth bark, and topped out at a hundred or so feet from the
ground. Which actually at the time seemed like a lot. But this was
something else altogether.

Long before what we now call Oregon or the United States were
named, this wild tree sprouted. The seed of the Discovery Tree was
nurtured within the scales of its cone. No bigger than the point of a
pencil, its seed was one of millions produced by its parent tree over

its lifetime. It had floated to earth on its brown wing, and unlike so many others, taken root on this forest floor, the perfect soft, moist medium. Its slender root had nurtured its lifting stem and grown to a trunk, straight and steady. In time the tree grew to dominate its grove, the biggest amid its neighbors, pushing through the canopy and pulsing with life through the years, century after century. Douglas fir is a long-distance runner that way; it doesn't really even begin to hit its stride until its second century. A native of the Pacific Northwest, it is a monarch of the Pacific Northwest woods, commonly living five hundred years, and the most venerable can be nearly three times that age. Over their lifetimes these grand old trees self-prune, dropping their branches from the bottom up, resulting in long, straight, branch-free trunks towering to a short crown with a wind-blasted top. The bark, as the tree ages, becomes more than a foot thick and deeply grooved and takes on a dark, rich, reddish-brown color. Its twigs are densely quilled with needles and the cones, two to four inches long, are perfectly symmetrical. They make fine food for animals, including chipmunks, mice, shrews, red squirrels, and songbirds that poach seeds right out of the cone. As lumber, old-growth Douglas fir is coveted for its strength, workability, and tight-grained, rosy gold beauty that sands to a silky finish. I write on a handmade desk of Doug fir.

As a living tree, it is the anchor species of the moist forests west of the Cascades. So this was royalty that I was about to encounter, a Douglas fir soaring more than twenty stories. Armored with thick plates of bark, the Discovery Tree glowered with gravitas. It has stood for some four centuries. And here stood I, hoping not to show that I was nervous. No. It was way worse than that. I was hoping to survive this encounter that loomed large as this tree in my tiny, short little marshmallow-soft mammalian life.

It was time to climb. I pushed down on the ascender latched to my foot and at the same time slid the climbing knot up the rope. I went up maybe a few inches. The top looked quite far away. Inch, inch. The first few moments were a wreck, my feet still on the ground and questing for the lift of the tree's branches to take over, but I didn't yet have enough height. I struggled, sweating. I looked over at Schulze—he was already up and climbing smoothly. A veteran of many climbs, he had worked for years in tree canopies of Central America; this was a climbing snack for him. His confidence, though, boosted mine. One more push of my foot and slide of the climbing knot up the rope and finally my feet were off the ground.

I careened into the trunk of the tree, and its solid, wooden rebuff hurt. But I'd come all the way from my home in Seattle for this, and the weather was predicted to shift to an afternoon thunderstorm. So this needed to start working. *Now.* What I needed was more momentum. Push, slide, push, slide, and keep at it. Finally I was clear of the swell of the tree's trunk at its base. Higher up, I hung free, truly in midair, the tree bearing my weight. I looked over my right shoulder. Yes. I was aloft. In the death zone, as climbers call it, above fifty feet, where any fall is likely fatal. Yet the occluding fear that took all my attention at the start had backed off a bit. I began to trust the rope and the harness. To see and feel where I was. The trunk before me was scaly and plated—a sign of the tree's great age. Above, the lattice of branches extending from the trunk beckoned. As I climbed, the distinct worlds of this one tree were revealed. The air smelled pungently of conifer needles and sap. Emerald moss clung to the trunk. The light was changing, brightening, as I climbed higher. Old man's beard lichen draped the branches, blowing in the wind, painting the air green. Fat juicy mats of moss topped the branches. A soft rain began to fall, more of a mist, cooling my face and silvering the tree's needles. Higher still, I got my wish: a broad branch perfect for sitting. I hitched myself over, perched like a bird, and looked out.

This was a totally different view of the forest. The canopy, seen from within. It was not the smooth dome it appeared to be from the ground, not at all. I noticed big gaps. Vast open space yawned all the way to the forest floor, where much smaller trees were getting their start. Gray spears of the broken tops of weathered cedars pierced the

Mark Schulze, director of the H. J. Andrews Experimental Forest, is an expert climber. His aplomb aloft boosted my confidence as we climbed this old-growth monarch, the Discovery Tree, at the Andrews.

canopy below me. I marveled at a sea of greens: the nearly black green of spruce, the malachite green of cedar and fir. The roughness of the canopy was such a surprise, with so much space between the tops, and the trees of such a variety of heights, ages, types, and sizes. This was no monoculture plantation, smoothly uniform as broccoli tops. Schulze and I paused for a bit and just hung out, gazing all around. He talked about one of his earlier jobs in his career, climbing in Central America—and all the ticks he had to pick off when he got home. So lucky were we, here in our Douglas fir redoubt, our treetop nation of two.

As I kept climbing, the branches were getting shorter and smaller in diameter, and the top more open, the view gaping to the soaring free air. I looked out into the open sky to distant mountains and finally relaxed fully, leaning back in the harness just for fun, enjoy-

For me, as a writer, climbing is an important way to experience the miracle of trees with greater intimacy.

ing its comfort and the spring of the rope. I dropped my head back, spinning in all directions. This was the raven's view, the realm where the tree's needles were combing the high winds. I felt as if I should be able to extend my arms and just float through the canopy's many greens. I thought about this tree's four hundred years, and of all it had seen and endured. Ice storms, wind, searing sun, lashing rain. But also the soft touch of bird feet, the silken glide of a flying squirrel, the first light of each dawn. I could hear Lookout Creek in the distance, purling through the old-growth forest. I know its shape, temperature, and clarity all were influenced by trees like the one I was dangling from, the forest shading the stream, keeping it cool, moderating the winds, the sun, and storm runoff. Needles and insects rained down from the canopy into the forest, nutritious food for the grazers and shredders in the streambed. Theirs was a tiny world of some of the smallest beings in this forest, but related and dependent in every way on the presence and life of the big trees that are the keystone of an old-growth forest. The flow of the wind and soft rain, the quality of the light, the temperature, moderate and sweet on my skin—all of these were influenced by these forest elders.

Schulze and I couldn't climb much higher, the branches and trunk were thinning, my rope topping out. It was time to come down. I paused on the way down as the rope whizzed through my hands. Schulze cautioned me not to go too fast—and anyway I wanted to take my time. So frightened at first, now I was reluctant to leave this tree. The gift of being held by it, the moments sitting and feeling the support of its ancient self, experiencing its community; I wanted to keep these memories with me. The way the wind and rain felt up here, the sound of the creek, heard from high above. With a thump, my feet were back on the ground. I unclipped from the rope and harness, bewildered at the abrupt change of worlds. The tree's motion stayed with me, the same way being at sea stays in the legs. I was still feeling the sky river of wind. As we packed up the gear, I tipped my head back and knew the tops of these trees would never look the same to me. I had sipped their rare air, felt the solid embrace of their branches, and savored their world. As a visitor, yes. But as a grateful celebrant too, of this tree, and a shared Earth that would not be as it is without old-growth trees like this one and the forests in which they abide.

2

WHAT IS AN
OLD-GROWTH FOREST?

THE WOODSTOVE TICKED with heat as Jerry Franklin settled in at the table by the window, overlooking the creek that rushes past his cabin off the grid in the Gifford Pinchot National Forest in southern Washington, 1.3 million acres extending along the western slopes of the Cascade Range to the Columbia River. Franklin has been visiting these woods since he was a boy. When this cabin on land leased from the Forest Service came up for sale, he didn't hesitate. He named it The Trees. Born Jerry Forest Franklin, perhaps it was inevitable that he would grow to be the world-renowned eminent forest ecologist he is. Franklin has been called the Father of New Forestry, the Guru of Old Growth. He's been an adviser to Congress and presidents, a director of the Andrews, and he has authored more than three hundred scientific papers. But through all that, he actually has had only one client. The trees.

"Come on, Trapper dog," Franklin said, and Trapper, his golden retriever, burst to his feet and headed to the door. It was time to go see the trees that Franklin told me have been his longtime counselors. Stepping out into the cool, wet June day, we rustled into our raincoats and walked down the dirt road pocked with puddles to a nearby campground, Trapper exuberantly taking the lead. We had not gone far on the woods road when Franklin headed into a grove of Douglas firs, towering in a circle, with a clearing in the middle. Franklin moved to the middle of the circle. He looked around at the trees surrounding him, like an embrace. It was here, Franklin said, that he accounted for himself, over the course of his career, reporting back to these old-growth firs.

Franklin's commitment to trees started early. He was thirteen or fourteen, playing in woods near his family home in Camas, which lies along the Washington side of the Columbia River, thinking about the forest being pushed farther back by development, the trees cut down and housing built. "And that troubled me, what was happening to it," Franklin said, recalling the moment his direction in life was set, in a promise he made in those woods. "I stood up and said, essentially, that what I'm going to do is, I'm going to spend my life doing whatever I can to cut the best deal I can for forests and trees in a world dominated by human beings. Now, I wasn't that sophisticated when I stood up in the woods and said it . . . it was something along the lines of, 'I'm going to do whatever I can. And I don't know how much that will be. But that's what I'm going to do.' And I said it out loud. That always makes a difference to me when you say things out loud. Even though there were only the trees to hear it. I declared myself." In his more than sixty-year career, Franklin told me, he has never wavered from that commitment. That career includes thirty years as a research forester for the US Forest Service, a professor of forest ecology at the University of Washington, and a director of the H. J. Andrews Experimental Forest, where he helped found the nation's Long Term Ecological Research Network (LTER).

Franklin grew up in the 1940s and 1950s in Camas as a pulp mill kid, where the high school sports teams still call themselves The Papermakers. Their logo is a hulking piece of papermaking equipment and their motto Mean Machine. Papermaking was king in his town along the Columbia River, where the abundant forests and river attracted developers of a paper mill in 1883, making newsprint for the Portland *Oregonian*. Through many name changes, ownership changes, upgrades, and expansions, the mill was the town's single largest employer and biggest source of local tax revenue. Everything in town revolved around the mill's ups and downs and many changes. It defined the waterfront, the city—even the air, thanks to the rotten egg smell of the mill's pulp-processing plant. This was the golden era in America of corporate beneficence, Franklin said, when the key industry in town was part of the community, giving money to the schools, gifting green space, investing in the place that provided the highly skilled unionized workers that made the industry profitable. As author Jamie Sayen so beautifully relates in *You Had a Job for Life* about company towns like Camas, workers earned good wages and benefits in jobs they could hold onto, until the next generation of their families would go to work in the mill. Franklin recalled:

> I grew up in an era in which most corporations felt a responsibility to their workers, and saw that as part of their responsibility, their corporate responsibility. And so things have evolved, it is very clear, that the capitalist system focuses on itself, and not on ordinary people and their needs. I grew up with the Crown Zellerbach corporation, and it was a good citizen and it did look after its employees and the communities it was part of. I have a real appreciation for the people that were doing the hard work; I never saw them as the real problem in terms of exploitation of forests. These communities and work forces are very, very important and need to be valued and nurtured. Of course today corporations and industry does not value them at all, and they are the ones that fall off the end of the table. But I really do identify with these workforces.

Franklin's dad worked in the pulp mill like his father before him. The family was too poor to have a car until Franklin was in middle school. Those camping trips to the Gifford Pinchot National Forest in the family car took him to the same campground where the trees stand today not far from his cabin. Franklin's dad would take him for a walk in the woods, the two of them lying down on the trail, while his father taught him how to listen, to tell the sounds that different species of trees made in the wind. An awkward kid, short-sighted, with no interest in sports, Franklin felt most comfortable in the forest.

Life changed when Franklin got to college. Spending his first year at a community college, Clark College in Vancouver, Washington, Franklin transferred to Oregon State University (OSU), earning bachelor's and master's degrees in forest management, and he then graduated from Washington State University, where he earned his PhD in botany and soils. One of the first in his family to go to college, he knew he would have to work to pay his way. "It was absolutely expected that you were going to have to work to get where you wanted to go, to get yourself through school, we didn't take out loans in those days. You had to either earn the money, get a scholarship, or both." For Franklin, work at the Camas paper mill was a lifeline. "It was a helluva place to make good money so I could go to college. The instant I could go to work in the paper mill when I turned eighteen, I did. It was challenging and it was rewarding because I was doing a man's work. And you always wonder about your ability to do things like that. Working in the mill was a very positive experience, I made a lot of money at it."

Franklin's job at the mill was to work the so-called extra-board—the pickup jobs needed each day to fill in wherever another hand was needed. Mostly he worked on the cleanup crew, shoveling bark from the debarking machine used to bash the bark off logs before they were chipped and cooked into pulp to make paper. "They go into a hydraulic barker to have all the bark knocked off, so you would have all along the way a lot of muck and bark and other pieces of wood," Franklin said. "It was a very dirty place and required constant cleanup. Sometimes things broke down, particularly when we were running black cottonwood, with stringy inner bark, and it would wrap its way around it. It was a place that had a lot of dirt and debris, and they had a crew that did nothing but clean up, with shovels, picks, rakes, and brooms. Common labor. I made $2.10 an hour. That was huge."

Beginning in the fall of 1954, Franklin put every extra hour he could into the mill, until the winter of 1956. He worked straight through both summers, and during the school year he got on a Greyhound bus from OSU on Friday afternoons in Corvallis, Oregon, and traveled to Camas to work two shifts over the weekend, going back Sunday nights. Staying at his family's house, Franklin worked Friday or Saturday nights or the graveyard shift into Sunday morning. Sometimes he could even get in three shifts on the weekend—"That gave me a really good income." When he got his first opportunity to take a job with the Forest Service, he had to think hard, it was such a cut in pay. Being from a mill town and raised by a mill worker shaped Franklin's views later in his career, when he was steering policies that would affect thousands of jobs in the woods and in mill towns. Places like Camas, or Forks, Washington. "We were poor, and I know what it is like to be poor," he said. "I have worked with a lot of different people that were working people. And had a lot of respect for them and the challenges of the lives that they lead. I come from a very common lower-middle-class background. I know at times environmentalists have thought of these people as the enemy, in a community like Roseburg, Oregon, or Darrington, Washington. I have never viewed them that way. In fact I have always felt that to a substantial degree, working-class people, blue collar people, are used by the system."

THE NEXT DAY, Franklin and I headed to the Cedar Flats Research Natural Area, just to experience it together. I had been exploring there on my own, but I wanted to see the old-growth forest there that I had already come to love through Franklin's eyes. As a college professor, he had taken a generation of students to Cedar Flats, to appreciate the old-growth characteristics he had spent his years defining—especially the special attributes of Douglas fir. Not that this was how he thought his career was going to go. "I was going to become the world expert in the subalpine forests, because people knew everything they needed to know about Douglas fir," Franklin said. "It turned out we didn't know *anything* about these forests other than how to cut them down and create plantations. And I have really come to appreciate the significance of the Douglas fir tree in making these forests what they are."

Widespread, broadly adapted to a range of conditions, and long-lived, Douglas fir persists for centuries and keeps right on persisting, even after it has died. By comparison most Western hemlock stands begin to fall apart after eighty to one hundred years and will not see a second century as an intact stand. Hemlock's strategy, Franklin explained, is in its ability to readily reproduce—not in resisting decay, wind, and insects. Its dead wood rapidly turns to mush. Meanwhile Douglas fir is just hitting its stride at one hundred years and only begins to take on old-growth characteristics at its second century. That means stands of Douglas fir can maintain their integrity and completeness for many centuries—several human lifetimes. Douglas fir is really the key to these forests, particularly the grand splendor of the old-growth forests.

For it isn't just big trees that matter, Franklin learned, it is big *old* trees, which play a different role in the forest. "Old trees are not simply bigger versions of young trees," Franklin said as we walked, Trapper zooming ahead on the trail at Cedar Flats, to find a good stick to gnaw. "If all you needed was big trees, we would be in better shape. But old trees take time." These big old trees gain not only size but the ability to host *processes* that take time—several centuries—to establish. "The reason they have such value is they offer such an array of niches for organisms, the old trees are just that much richer in the variety of habitats and conditions they provide," Franklin said. "The old trees are going to have much more of the decay, that resistant heart that decays so slowly. It isn't just a matter of size, it is a matter of the tree evolving over time into much more individualistic and niche circumstances, and continuing to provide very specialized habitats once they die."

Like people, each tree's life is an adventure, and the older it is, the more things happen to it, all of which are recorded in the shape of the tree. Also like people, trees accumulate unique features from their injuries, infections over the course of their life, and their responses to them—the bends in their trunk, their reiterated broken tops, the shapes in their growth, and the reach of their branches. Trees respond to changes in light conditions too. They surge into gaps and sprout epicormic branches to capture the light. These are branches that grow from buds in the trunk that don't sprout unless some bonanza of sunlight stimulates their growth—such as a tree next door falling over. Across their long seasons of time, big old trees amass larger branches that host many lives. A single stand of old-growth trees can nurture more than fifteen hundred species, as Franklin and his collaborators found in their 1981 paper on old-growth ecosystems, including many invertebrates and species that spend their entire lives in the canopy. The canopy includes several insect habitats, both vertically and horizontally, with some found only in the upper canopy, others in the lower branches, some on major limbs, and others among twigs and foliage. The northern spotted owl and northern flying squirrel are some of the best known animals that thrive in old-growth tree canopies. But there are others, such as the unsung tree vole, that are just as devoted denizens of old-growth, raising multiple generations in the canopy of the same tree.

Almost every surface of an old-growth tree is occupied by epiphytic plants that take their nourishment from the air. More than one hundred species of mosses and lichens make their home in old-growth trees. The dry weight of mosses and lichens on a single old-growth Douglas fir can weigh as much as sixty-six pounds, not even counting the crust-forming lichens frosting the bark. Nearly half the total weight of epiphytes typically is the single leafy lichen *Lobaria oregana* growing mostly on the upper sides of branches and twigs. They help form the soil on the tops of the branches, with the soluble minerals they take from the water flowing over them, the dust and litter they trap, including needles, all accumulating year by slow year. Lichens, especially *Lobaria oregana*, feed the forest with the nitrogen they fix, ultimately made available to the whole forest through leaching, litterfall, and decomposition—as much as nearly five pounds per acre per year. This significant gift of food manufactured by lichen is nourishment largely found only in old-growth forests. For sheer photosynthetic capacity, nothing beats the scale of a massive old-growth tree—just one of which can have more than 60 million individual needles with a cumulative weight of more than 440 pounds and a surface area of 30,000 square feet.

Old-growth forests have that distinct look: broken and dead tops and branches, big gaps in the canopy, and raggedy bottle-brush branches; thick, plated, torn, and streaming bark—these characteristics all spell age and history—centuries of it. "As you age you become decadent, we can relate to that as human beings, with trees the decadence is what creates the richness and complexity," Franklin said. "The old trees develop very complex canopies with dead tops and new, reiterated tops, and big epicormic branch systems. And they have a lot of rot and decadence in them, which is important to wildlife. The number of microniches associated with older trees, as opposed to younger trees, is immense. Just think about all the cavities, and the big branches, and the Douglas fir is quite extraordinary, because it lives so long."

As he says this I tip my head back, looking up into the tops of a big Douglas fir. To be in the presence of a tree this old—at least four hundred years—is a quieting experience, the sort of moment the word *awe* was made for. Far up its massive trunk were the epicormic branches that are so special—branches that sprout from the trunk even after it has self-pruned its lower branches. These epicormic branches have a different shape and come only in later life. They can even wrap partly around the tree, seeking sun, and have a broad, flat, fanlike shape. Just perfect for growing dense mats of moss, so deep and long-lived that they make their own soil. These are the nesting platforms so sought by birds that don't make their own nests, such as the marbled murrelet, an endangered species listed for federal protection under the Endangered Species Act.

For all their big trees, an old-growth forest is not everywhere a dark, gloomy place. They are characterized by structural complexity, not only in the trees themselves but in the forest they make up, full of gaps between individual trees and larger disturbed areas that provide the crucial sunny open areas where the next generation rises. These canopy gaps are a signature of the old-growth forest, where all the stages of life and structures of a forest are present. These stages and structures all matter, Franklin said, and are a feature of naturally regenerated forests. This beautiful mosaic is diverse in its species, structure, and life stages. And old as they are, even the oldest trees are relatively new, regrown after stand-clearing fires centuries ago. Nothing is as old as the land itself and its endless cycles of life and death, destruction and renewal. Perhaps it is seeing this in the uneven

This marvelous fructifying stump is typical of the glory of the Cedar Flats Research Natural Area outside Cougar, Washington.

ages and conditions of these old forests that makes them so moving, speaking without words of our own mortality and inevitable transition and continuity to what and who will come after us.

There is a rightness in this, a comfort that only the language of old trees can speak. Old trees matter. It is here that wildlife finds refuge, in the nooks, crannies, and caved-in nest cavities and cool ferny glades of their understory. The thick, complex, and furrowed bark of older trees also creates the homes for insects that in turn feed birds, bats, flying squirrels, and other species. This assemblage of ages and species can be created only over centuries. Age and diversity also make these forests resilient to fire, insects, and drought. In dry summers of the Pacific Northwest, old trees are capable of pulling huge amounts of water from deep in the soil, shared between their roots.

It was Franklin and a team of collaborators based at the Andrews who revealed the wonders of old-growth forests—*considered biological deserts at the time.* Their foundational research paper would become the basis for understanding these forests, and ultimately saving them. At the Andrews, Franklin developed not only one of his most important bodies of research but his way of thinking and doing research. Instead of lab experiments picking apart pieces and parts, he learned by *looking* at the entire forest as an interconnected landscape and ecosystem. "I always felt intuitively that the old-growth forest really held a lot of knowledge. I always knew even when I was a student at forestry school that there is more to this than meets the eye, and more to this than anyone is telling me, because we didn't know anything," Franklin said. "I had a sense there was more to it; for me, my constant reference was my intention to do the best I could for the forest and the trees. And that meant getting knowledge relevant to clarifying the importance of these forests, so that is what I did." It was the collaborative work of a team of scientists, working to piece together the story of the forest and how it functioned, that made the magic happen.

"To me, my strength has been my ability to put pieces together into a larger whole, that is the role I play," Franklin said. "Certainly when I

Jerry Franklin talks with me about the old-growth characteristics of this forest at the Cedar Flats Research Natural Area. Photo by Doug MacDonald.

was doing the research at the Andrews I was thinking constantly about where the gaps were, where we needed more information in order to get a more complete picture of that system. One of the things I also liked to do is get people together to brainstorm." That is the origin of that now famous 1981 paper, of which Franklin is the lead author, "Ecological Characteristics of Old-Growth Douglas-Fir Forests." It was the groundbreaking result of a deep look at the structure and function of an old-growth forest, by a team of scientists willing to be instructed by the forest itself. What was all around them in the old-growth forests and streams at the Andrews, they realized, was dead and downed wood. What was its function? With so much of it, everywhere, it had to be important. But how? It was right under their noses, yet no one knew anything about it. "There emerged this whole question of dead wood," Franklin said.

That was a really important lesson, "Oh my God, dead wood is important." Once you were ready to accept that statement, you could see the many ways it was important and then it was, "Oh my gosh, there is different kinds of it out there, and we need to know how long that stuff lasts." That is probably the first really big thing, and it is reflected in that 1981 paper, the importance of large live trees and large downed logs. Once you began to say, "Let's think about why this is important," then you tabulate all the organisms that utilize this as habitat, it becomes a part of wildlife habitat guidelines. Small mammals had been viewed as problems rather than inhabitants. We were asking, "Who is using these logs and what are they using that for?" We were trying with dead wood to understand more of its dynamics, to get a sense of the rate of decay.

No one had paid any systematized scientific attention to this before Franklin and his colleagues asked these questions, but dead wood, it turned out, is crucial to the life and function of an old-growth forest. Here was a spectacular diversity of rot, a marvelous mortuary of cellulose crucial to the life of the forest, and remarkably persistent.

Big old Douglas fir snags—dead trees or parts of trees—can typically stand some fifty to seventy-five years before being reduced to stubs less than thirty-five feet high, while western redcedar, dead as dead can be, nonetheless will typically remain whole and standing as long as 125 years. Large standing dead trees are the most valuable to wildlife, prized by hole-nesting birds. These large snags are the result of large trees, so they are the special gift to the future of old-growth forests. Again time is the key. It takes about 150 years for a natural stand to develop snags even twenty inches in diameter. Douglas fir snags typically disintegrate from the top down. The top and bark are the first to go, then the trunk finally calves off nice big chunks to the ground, leaving a short snag or juicy stub. Western redcedar tends to shed its bark, standing gray and bark-free and entire until its roots decompose and the snag falls.

One of the most important functions of all this dead wood—from logs on the ground to a standing dead snag—is as habitat for dozens of species of wildlife, from birds to bears. The more snags and cavities, the more wildlife a forest can support. Big hard snags that the primary diggers, such as pileated woodpeckers, will excavate are especially important. The old-growth forests—with all their stages of decay, diversity of life stages, and species—meet a panoply of needs, from hibernation to nesting, roosting, resting, dating, and feeding. It is dead wood that nurtures new life. Once a tree dies, its next life is just beginning. Once fallen to the ground, a tree has a whole new purpose: it has not died at all but attained a new stage and importance. Messy masses of large woody debris all over the ground are a glorious signature of a wild old-growth forest, tangled, jack-strawed, and uninterrupted in their conversation with the legacy of the forest's past generations. These legacies give rise to its future; the forest lives on with continuity to the *next* generation.

SO MUCH DEAD WOOD! It's not unusual for as much as eighty-five tons *per acre* of dead and downed logs to be piled and heaped over a twenty-five-acre watershed forested in old-growth Douglas fir and hemlock—and some watersheds would be plastered with much more. These logs are the bursting food pantries of the forest, replete with nitrogen and phosphorous, key nutrients for plant growth. Rotting logs hold large amounts of water, which makes them important as rooting sites for tree seedlings and sites for nitrogen fixation. The stability of these nurse logs is part of what makes dead and downed logs so important in an old-growth forest. They make primo wildlife lookouts and are perfect spots for feeding and reproduction. Dead wood offers protection, cover, bedding, and swell places to store food.

Amphibians love a nice moist rotting log. The longevity of big old logs also makes them an important refuge in major disturbances, allowing animal populations to repopulate by providing pathways along which small mammals can travel and on which seedlings of new life may sprout.

Dead and downed wood also is home to the mycorrhizal fungi that provide the associations that many tree seedlings need to get their start and to thrive. Western hemlock in particular roots its new seedlings in nurse logs, and so does Douglas fir. This is why in any old-growth forest, colonnades of standing trees often will show a barely visible hump of a long-utilized log at their base, nearly dissolved with time: the nurse log on which they all sprouted. Seedlings are both more numerous and taller on nurse logs, for the moisture, nutrition, and competitive advantage the logs offer seedlings in getting up above the forest floor. Dead and downed logs also are important to humans, safely locking up for even more centuries the carbon these trees pulled out of the atmosphere when they were living, helping to moderate climate chaos.

Dead and downed wood is just as important in water. Big wood enters streams by blowdown, slides, and avalanches, but streams will also help themselves to a feast of wood, undercutting stream banks and toppling trees into the current. When the big floods come, they move wood downstream, where it builds logjams. Logs force the main channel to avulse, dig new braided side channels, and excavate pools that are prime habitat for fish. Wood and water, when they are allowed to, are always on the move, building, tearing down, and rebuilding, in an unending dance of forest and stream. But it takes big wood to provide the structure that makes it all work. This accumulated woody debris provides the cover, stability, and homes for everything from the tiny insects and crustaceans, the shredders and grazers and scrapers at the base of the food chain, feasting on forest detritus in the stream bottom, to the snails and juicy caddis flies and fish that devour these nutritious beasties. Woody debris also provides crucial corridors for wildlife, zooming across streams, and is great for lurking, perching, and feeding.

So important is dead and downed wood in the life of forests and streams, some scientists have built their whole career on it. Mark Harmon of OSU—locally known as Dr. Death—has a forest decomposition experiment underway at the H. J. Andrews Experimental Forest intended to carry on long after he is gone. Initiated in 1985, the experiment on log decomposition is designed to test the effects that species and size of logs have on decomposition and nutrient cycling. Setting up the experiment involved cutting 530 logs of four species—Douglas fir, Pacific silver fir, western hemlock, and western redcedar—bucking them into twenty-foot logs, putting in them in the forest in six sites, with eighty to ninety logs per site, and just letting them sit. Harmon and his collaborators sampled everything along the way: food web structure, nutrient cycling, insect activity, moisture content and temperature, how much bark remained, moss and lichen cover, and more, checking in on the logs every year. They cut slices from the logs and hauled them back to the lab, to examine their heartwood and sapwood and to assess their density, moisture, and nutrients. They studied respiration from the logs, exuded by their long slow decomposition. The very breath of rot.

I met up with Harmon at the Andrews in the summer of 2022, to go have a look at one of his log decomposition sites. He looked as if cast for the part of forest decomp guy, with the shoulders of his T-shirt dusted with hemlock needles, a wild beard, and exuberant interest in all things dead. After a short hike, there we were. The site would have been easy to walk by; it's basically a clearing with a casting about of big dead and downed logs, with some telltale sciencey-looking things often seen in the Andrews—a funnel here, a plastic tube set in a log there. What has already come out of this low-tech experiment has been surprising. The sheer number of lives these logs harbor was not expected, nor was the revelation of how quickly life moves in right after death, with ambrosia beetles bringing in the fungus that its

Forest decomposition expert Mark Harmon pokes his finger into the soft wood of a decaying snag at the H. J. Andrews Experimental Forest. Dead wood isn't so very dead at all: it teems with decomposers, feeds birds and bugs, and houses everything from bats to bears.

young eat. Mites, nematodes, paramecia, tardigrades—this dead wood was alive. "The diversity that came on in just the first few months was astounding," Harmon said. "There were dozens of insect species that were boring in, laying eggs, other insects were eating those, it was amazing how fast life came back to the log." The logs were sponges, with 25 percent of the water that fell on them stored and slowly evaporated. The water coming out of the logs was enriched with ten to twenty times the concentration of carbon and nitrogen, making them a source of food for the forest. Teeming biodiversity flocked to the logs, with fungal fruiting bodies and mushrooms thriving in the rot.

Harmon and I passed a handsome standing snag, with a cascade of cinnamon-colored chunks heaped all around its base. "That live tree there?" Harmon said, pointing to a tree next to the snag. "It's even deader than this dead thing here, by the proportion of living cells in an organism. We have this so-called dead tree here, well, some studies have shown a 10 to 20 percent living mass, in fungi and decomposers." Yet a living tree is mostly dead, but for the leaves and cambium in sapwood, the roots and mycorrhizae connecting them. "You take all those parts and figure it is mostly 5 percent alive. So something 20 percent alive is dead? And something that's 5 percent living is alive? I'm a little confused here. It limits thinking to say, 'That is alive and that's dead.' They are alive in different ways. We look at it this way: it's dead, it has no value, but what about all these other things?"

Decomposition for Harmon has been a fifty-year obsession. Tree species decay at different rates. It can be more than four hundred years before a cedar log completely decomposes, but a silver fir is gone in sixty. Doug fir decays at half the speed of hemlock. *They are alive in different ways.* Harmon pointed with his finger to the infinite galleries tunneled into a Doug fir log, the decay organisms revealing its structure, now soft enough to pull apart with my hands. It still had some umph to it, though, even where it was rotting. Not so the hemlock nearby in a pile of mush like a spitball. "That's classic white rot," Harmon said. "Eventually, it will disappear." He marvels at the revolution underway in recognition of the importance of woody debris, from when foresters piled up any unmerchantable wood as trash for burning or dumping. "In the eighties you would just see piles of rotting logs by the (logging) roads." Taking all that away was robbing the soil of the nutrients, moisture, and fungi in the wood. "If you take all that away, of course you impact how the system functions," Harmon said. Yet many scientists could not see the importance of dead and downed wood. To Harmon, this was a problem not only with understanding dead and downed wood but incurious, timid inquiry.

It's "Wait a minute. We have not been seeing this, but it's everywhere, let's look at it," rather than finding out more and more about less and less. That is not a good path for science. You have to ask, "What are we missing?" Scientists had a huge obsession with net productivity—how much carbon forests were sequestering through the photosynthesis of living trees. But little attention was paid to the next phase of the forest. There is another stage, you have to go into this next stage, the afterlife part, if you want to figure out the system.

Harmon and Franklin were a good fit that way, being open to discovery. "What attracted me to Jerry is intuitively we keep coming back to the part, 'What am I missing?' What we are really talking about is the process of discovery," Harmon said. "A lot of scientists are just dialing it in, just going through the steps. There is no creativity, it's all about the stats and my God it is so boring and pointless. You are just following a formula. But not everyone can be a grand thinker, a grand synthesizer. That is the point of a team, and one of the things I learned." That practice of long-term, place-based inquiry, by collaborative teams working across disciplines, became a lifelong signature of Franklin's style and particular genius. "If you are about discovery you have to be open to discovery, to be willing to make mistakes, willing to break rules or at least not be so mired in them that you are self-editing," Harmon said. "Observe. Don't be so judgmental. Don't be afraid that you may be wrong. As a scientist, if you think you know it all, you are even wrong about that."

Leading and participating in teams was central to Franklin's career, and his instincts in leading bands of young and senior scientists in open-ended inquiry played a crucial role in developing the science now widely accepted of old-growth forests and how they function. Franklin's genial, collaborative style became a brand and a method. He called these gatherings "pulses," and the freewheeling work that came out of these cross-disciplinary happenings—whether camped in a downpour in the mossy old-growth of the Hoh Rain Forest in the Olympic Peninsula or gathered at the crater of Mount St. Helens—continued throughout his career. This style of inquiry inspirited the purpose and method of the Long Term Ecological Research Network that Franklin helped found. This was scientific inquiry inspired by reading the landscape, asking questions, building collaborations, and sparking brilliance in actually seeing and understanding the obvious—such as dead and downed wood—that no one had bothered to see, much less study, before.

The revelations that came out of this early work about Pacific Northwest old-growth forests—that there is such a thing, that it has value, especially all that dead and downed wood—was not particularly welcome at the time. That these old-growth forests are neither decadent, unproductive ecosystems nor the biological deserts they were thought to be was not appreciated by professional foresters bent on cutting them down. Yet here came science that found these forests replete with compositional, functional, and structural features importantly different from the "thrifty young stands" industry favored. All the elements that set old-growth forests apart were, inconveniently, crucial, interrelated, and irreplaceable—and the product of *time*, lots and lots of time. Large live trees, large dead snags, large logs on land, and large logs in streams could only be the product of *time*. Centuries.

This understanding threatened standard clear-cut logging, high grading the biggest trees, bulldozing every bit of the dead and downed wood and slash into a pile and burning it, then spraying herbicide on the ground to kill off any understory before planting a monoculture of seedling trees for another quick rotation to the next cut. Everything Franklin and his collaborators had published in that 1981 paper showed the poverty that would result in a natural forest transformed in that way, with every biological legacy, every diversity of species and age class and interaction eliminated. The implications of this science, both for forestry and wildlife policy, were enormous—and not popular with industry. This work and its method—basically looking at the landscape and working to understand how it functions—also was frowned upon in academic and scientific circles, where scientific questions more typically were launched from hypotheses and proven in lab experiments. As Franklin put it:

> We learned by looking at the old forest. That is how we were taught. And the richness and the complexity of it also taught us to be humble. It was a tremendous object lesson in humility and respect, as well as knowledge. The lessons, in terms of knowledge, but also lessons in terms of you have to be pretty arrogant to just assume this is nothing more than a source of wood and you can essentially duplicate it by planting rows of trees. We took a lot of shit from ecologists that thought this was just a hobby thing. We had a lot of negative feedback on the dead wood work, attacking it as just hobby stuff that had no relevance to science. A place like Harvard would have nothing to do with it, it wasn't the sexy sort of stuff, the hypothesis testing the academic scientists are so excited about. . . .
>
> I was often angry at the eastern ecological community, the ones writing all the papers, writing all the text books. What they were writing wasn't taking account of the forests that I was in. And I was bound and determined that one of the things I was going to do was to see these theoreticians have to take account of these forests. We established it was an important structural

Industrial clear-cuts like this one impoverish the landscape, replacing natural forests with monoculture fiber farms.

element in the forest. It has to do with mortality of *trees*, not just leaf fall or branch fall. That means an investment of time.

This lack of respect for *time* is what's wrong with industrial clear-cuts, Franklin found—the plantation-style forestry that robs everything the old trees had taught him was important. The herbicide that kills the young growing understory that feeds the soil, the replanting with a monoculture that will never have the complexity of structure, age, or species of a true natural forest. These conversions of natural old forests to plantations were stealing from the future; these once naturally-forested sites were now croplands that could never become the old-growth of tomorrow. "I refer to plantations as one of forestry's great deceptions," Franklin said. "We were fooled. People have to realize that all is not well. Foresters deceived themselves into thinking plantations are forests, and they are not. These are great for the stockholders, but nobody else."

The importance of old and mature forests is even more evident as human-caused climate warming, stoked by the burning of fossil fuels, menaces the planet. Forests take up about a third of the human-produced carbon dioxide emissions in the atmosphere. Trees take this carbon into their tissues as food, through the process of photosynthesis, and store the carbon in their wood, leaves, and roots. Big trees—both mature forests of more than one hundred years and old-growth at two hundred years and older—are the largest and most persistent sink of carbon. No climate strategy is as cheap, reliable, or effective in reducing carbon in the atmosphere and storing it long-term as simply letting trees grow. Scientist Beverly Law at Oregon State University and her coauthors established this in a 2022 paper. Law and other scientists, including Franklin, have called for the preservation of all old-growth and mature, naturally-regenerated forests everywhere. Forests are such complex systems, Franklin says, that humans can't even know what we are losing as we continue to cut old and mature natural forests. The first step for biodiversity, climate stability, and human well-being, he says, is to reserve all existing naturally developed mature and old forests. They are irreplaceable, certainly in our human time scale, and probably never at all.

Franklin remains hopeful people will see it is in their own best interest to preserve and protect the forests that preserve and protect us. Surely, by now, the science is clear on how forest ecosystems work, how to care for them, and their crucial role in our world. "Doesn't it matter that we know more?" Franklin said. But, he cautioned, it doesn't matter *what* we know if we ignore it and just go for the quickest return on the dollar. "How do you survive on a planet where the dominate economic system is return on capital?" he asked. "And the belief that no natural system can't be substituted for efficiency and maximum return on capital? As long as that is the dominant economic system, we are in big trouble." *All* forests must be managed for ecological benefit, not just monetary return, Franklin contends, including industrial forest lands. It takes longer rotations between cutting. Far greater retention of trees even on the harvested land. And very limited use, if any, of herbicides on the land, to allow natural regeneration of a diversity of species. "In the debates over forests, stakeholders have been led to believe that there are only two alternatives—tree farms or preserves. Nothing could be further from the truth," Franklin said. "Forests can be managed simultaneously for environmental, cultural, and economic values." There are no throwaways: All forests should be managed with considerations for biodiversity and other important functions, Franklin said.

This is an approach Franklin calls ecological forest management, rooted in an understanding of forest ecosystem dynamics and processes—instead of forests as simply a supply for merchantable wood. This approach to forestry is intended to protect even forests that are cut for timber as living ecologies—with the preservation of the soil, dead and downed wood, snags, and large swaths of standing trees left for wildlife and seed dispersal (not just pathetic sad clumps and individual leave trees that quicky blow down). This is *forestry* (not tree farming), and it has more than economic return in mind—forestry in which the forest's capacity for renewal is retained. "The most productive lands do good for society, not just Wall Street," Franklin says. He is a soft-spoken man, but when he says this, he seems with the emphasis of his conviction to be shouting. "I am not going to shut up about it. Society has been led to believe that you have to have one or the other—tree farms or preserves. Nothing could be farther from the truth. Forests can be managed simultaneously for environmental, cultural, and economic values. All forests should be managed with consideration for biodiversity and other important functions. Islands of habitat in a sea of not-habitat is not going to work. We need to be preserving biodiversity everywhere. I doubt anybody has ever drawn the boundaries on more reserves than I did. But you don't give up on the rest of it."

3

DRAWING THE LINE

IT WAS ACTUALLY A VOLCANO that taught Jerry Franklin some of his most important lessons about the poverty of industrial clear-cuts, and just exactly what is wrong with this type of forestry. It took an even bigger disturbance than any logging show for him to get the point: the eruption of Mount St. Helens, on May 18, 1980. After the southwestern Washington volcanic peak spectacularly blew its stack, Franklin was one of the first in a team of scientists to chopper into the blast zone. He already had a pretty well-worked-out expectation in his head of what he would see: Death. Destruction. A moonscape. The research team landed just ten days after the eruption. "I was overwhelmed by the scale of the destruction," Franklin said. "That was my first impression. But my second impression was there is a lot going on there. I stepped out of the helicopter and looked down, and here was this fireweed shoot coming up through the ash, and all kinds of things crawling around, ants and beetles, all kinds of activity, it was 'Oh, this is not what we thought.' It was just a lot of fun. I love epiphanies, we think we know . . . and then *whack*, okay we were totally wrong, what brilliant scientists we are. I love that kind of thing."

Everywhere there were important biological legacies from before the blast. Pocket gophers that had waited out the eruption in their burrows. Shoots and seeds and sprouts and mycelia that quickly grew through the ash layer. Animals that migrated into the blast zone, from red-legged frogs in the nearby forest to wild steelhead that repopulated the Toutle River that had been plugged with sediment and ash. "St. Helen's really punched us in the face. We were just dumbfounded at the diversity of the legacies left behind, both living and dead,"

Julia Jones, professor of geography at Oregon State University, pioneered research that showed forest plantations established after clear-cutting deplete summertime streamflows. Photo by Doug MacDonald.

Franklin said. "Once we had that wake up, we looked at all the other kinds of disturbances, like windstorms and fire and insect outbreaks, and realized, hey, all these things have incredible legacies. And the contrast with what we were doing with forestry, clear-cutting, was totally in opposition, to cut a forest and start over." The still unfolding discoveries made on Mount St. Helens, one of several volcanic peaks in the Cascade Range of the Pacific Northwest, detonated a whole new theory of what ecologists call disturbance. What Franklin and his cohort learned was that biological legacies were the essential spark of new life and continuity for the next generation. The lesson was clear: with enough of these legacies still intact, a landscape could surge back to life even after a volcanic blast.

That lesson—and what Franklin already had been learning in the old-growth forests of the Pacific Northwest about the integrity and completeness of its vital forces—got him thinking. Clear-cutting was a method of forestry that removed not only the merchantable wood but every bit of the natural forest—the logs, the bark, the branches, everything but the stumps. The forest floor was compacted and churned by heavy equipment, the hosts for its mycorrhizal fungi removed along with the trees, and the soil sprayed with herbicide to suppress the understory plants. All these important biological legacies were removed. Next a monoculture of species was planted, in tree seedlings all the same size and age. Everything that Franklin and his team had learned from the mountain about how a landscape resets to new life was being destroyed by deliberately suppressing and removing the biological legacies of the forest. A living natural forest was converted to industrial agriculture, a farm for growing wood.

Clear-cutting creates a big and lasting effect, changing the hydrology of entire watersheds. In 2016 scientists Timothy Perry and Julia Jones at Oregon State University published their findings that average daily summer streamflow in basins with 34- to 43-year-old plantations of Douglas fir was a whopping 50 percent lower than in reference basins with 150- to 500-year-old trees. Lead researcher Catalina Segura and her coauthors from Oregon State University reported in a 2020 paper that these effects not only persisted but held true even when plantations represented only a few percentages of the watershed area—and the presence of riparian buffer zones made no difference. This is because in the clear-cut, the young, fast-growing trees have far greater sapwood area and higher concentrations of leaf area in the upper canopy. They have less ability to control their transpiration and a higher rate of evapotranspiration. So they suck up water at a far higher rate in summer, just when it is naturally scare, compared with big old trees. This legacy of clear-cutting is surely only to be intensified when compounded with the effects of climate change, including lower summer streamflows and higher summer temperatures.

The clear-cuts studied did open areas favored by some wildlife, but they removed the cover, food, and shelter other animals absolutely require—including the northern spotted owl, an animal that would come to figure large in the future of these Pacific Northwest forests. Cutting these Douglas firs at sixty and forty years, as is typical in industrial rotations, means these trees will never come into their full ecological capacity in any way, whether for carbon sequestration, habitat, nurturing the soil, or persisting as legacies into the next generation. Spraying herbicide on the cutover ground also means there is no diverse understory, with its elderberry, its blackcaps, its foamflower—all wonderful food for pollinators and birds and small mammals of every sort that feed the raptors and shelter a whole catenation of life. Herbicide cheats an entire crucial stage in the life of the forest: when the canopy is eliminated, water and light and nutrients are capitalized on by herbs and shrubs in an exploding diversity of plant life, a natural cornucopia feeding the soil and wildlife. These early communities after a major disturbance are crucial to the mosaic of a natural forest that includes multiple life stages.

Gaps in the canopy provide the opportunity for new life that resets the forests and allows continuity for the next generation. "You have a lot of food to be eaten there, and you develop an incredible community of animals of all kinds, an incredible diversity of food webs," Franklin said. "For me the epiphany was the realization that those periods between a disturbance and when a forest canopy reestablishes is the most biodiverse period in our forest landscape. It is really important in a natural forest, because once the forest regains control [closes its canopy] it will be centuries before there is another opportunity for that." It is the total suite of conditions and stages of a natural forest that makes it so biodiverse, so resilient, so wonderfully alive: the new open areas regrowing, the dead and down wood, the old-growth, an understory, and the life of the canopy. All of these together are what make a forest—not a crop—through years of continual change effected by windstorms, fires, ice, and drought. It is a complete forest—not just old-growth itself—that is a biodiverse forest and a resilient forest, a true living, ever-changing forest. Not a crop and not a museum. Everything about the way the land was being managed on industrial lands, converting a real forest to a tree farm, was the opposite of those natural processes.

Unfortunately these lowland Cascade forestlands were among the first to be converted to industrial plantations because they are among the most productive in the Pacific Northwest. Changes in the tax laws in the 1980s that led to changes in landownership to real estate investment trusts (REITs) had drastic implications for the management of lands such as these. "Things changed," Franklin said, looking over cutover land on the mountain. "What is being done on corporate lands is fiber farming with no regard other than what is legally required for any other value. When I see something like this I am really sad. Forestlands should not be treated in this way. This is one of those situations where return on capital does not produce the best return for society. This is like converting a forest to a cornfield. Or a wheatfield. And I don't like having farmland in the middle of the mountains." Both the present and the future are impoverished by such management, Franklin said.

Clear-cutting doesn't do well at all in terms of providing conditions for development of a lot of that biodiversity associated with early successional ecosystems. First off, we try to eliminate any legacies, and we try to get back a forest canopy as quickly as we can. And spray herbicide so the trees can grow back faster, doing everything possible to terminate the forest-free period as fast as possible. The owners of the corporate lands are not really doing forestry at all anymore, they are doing fiber farming, and the bottom line is to have a high return on what they describe as capital. They are part of a global market, and they are growing trees as capital and you can't grow a plantation for very long. They are constantly accumulating annual cost. And the increasing value of what is out there forces you to cut at maybe thirty, thirty-five, maybe forty years. It doesn't go out any further than that. They are not any more trying to grow timber, the object is to grow wood to provide a return on capital. That is very different from the old days, when they actually were trying to grow a lot of wood because they were supplying their mills. Well, now those mills are not associated with those timberlands, and the goal is not to provide raw material. The goal is to provide a high return on capital by growing trees in plantations with short rotations.

Such management, Franklin said, does not realize either the potential of the species—the magnificent Douglas fir—or the land. "It's wasteful. These are the most productive forestlands in the world, and they are being wasted in terms of wood production and carbon storage. These are not functioning forest ecosystems. These are tree farms for growing wood to produce a high return on capital. All other values have been sacrificed—for the watershed, the wildlife, the climate. That's only a good thing if you are a Wall Street investor." In addition, these cutover lands are not resilient to fire, they suck water like mad, and they do not even produce wood as well as a forest, being cut so early in the life of a Douglas fir. The best solution for these lands so hard used at this point, Franklin said, is to buy them and try to restore their integrity. "The best nice thing about REITs is, if you have enough money, they will sell you anything," Franklin said. It's just all about money.

THESE LOWLAND CASCADE FORESTLANDS converted to industrial plantations have enormous potential for wildlife, carbon storage, and watershed protection. Only time can heal an industrial clear-cut: carbon storage is set back considerably and does not approach old-growth storage capacity for at least two hundred years, Franklin and his colleagues, including decomposition expert Mark Harmon, re-

Industrial clear-cuts today comprise much of the private forestland in the lowland Cascades. Largely owned and controlled by distant investors, these are not forests but fiber farms, managed for maximum efficiency and profit. Photo by Steve Ringman, *Seattle Times*.

ported in a 1990 paper. It is clear-cuts like these that spurred profound change in the woods on the US side of the US–Canada border, in the adoption of the Northwest Forest Plan in 1994. It was not exactly a tree but a bird—the northern spotted owl, whose old-growth habitat was threatened by clear-cut logging—that swung open the door to litigation that led to adoption of one of the world's most ambitious multispecies forest management plans. The plan pushed through by the Clinton administration and approved by a federal judge shut down harvest on millions of acres of old and mature forest on federal lands, from Washington to Northern California.

Franklin well remembers seeing his first northern spotted owl. He was hiking the H. J. Andrews Experimental Forest on a warm afternoon, and right next to the trail about six feet aboveground was a roosting pair, snugged in a vine maple. "I sat down across from them and just watched and listened to them, and finally, I had to get up and leave," he recalled. "I didn't even know what they were at the time, I was just a young forester, I had never seen a spotted owl, they were just a very gentle creature." But it was not long before the owl would make history. As clear-cut logging chewed up the old-growth forests of the Pacific Northwest, opponents filed a lawsuit in federal court, leveraging a convergence of new ecosystem science and environmental laws passed in the 1960s and 1970s, including the National Environmental Policy Act and the Endangered Species Act. Opponents argued that decades of unsustainable clear-cutting in the national forests was driving the spotted owl, the marbled murrelet, and Pacific salmon to extinction. US District Court Judge William Dwyer issued a court order stopping harvest on federal lands in Washington, Oregon, and Northern California in 1991.

Against all political advice, newly elected President Bill Clinton waded into the region's biggest natural resource conflict. In the late winter of 1993 he convened a Northwest Forest Conference in Portland, Oregon, which he and Vice President Al Gore attended. Gathered around a large conference table were Franklin and other top scientists, along with mayors, mill owners, and the president and vice president. The mayor of Hoquiam wept as she spoke of her town's plight, so dependent was it on logging. I was there, covering the meeting for the *Spokesman-Review* (Spokane, Washington), and I still remember it. Here was the president of the United States, seated in a talking circle with people from humble logging towns. It felt like just the sort of thing a president ought to do. Clinton impaneled a team of top scientists, including Franklin, to come up with a solution to the war in the woods. Their charge was to generate a scientifically credible, legally sufficient forest plan that would satisfy the judge, allow some harvest, and conserve multiple species, not only the northern spotted owl but salmon and other animals that depend on a functioning, ecologically sound forest.

In an unprecedented effort, Franklin and other experts assembled a team of field scientists at a hotel in Portland—eventually hundreds of specialists would be involved in the effort—and literally sketched out the future of the region's old-growth forests, drawing lines on maps to set aside the places regarded as most important to species survival. There were no lobbyists, no outsiders, Franklin recalled, just the people who knew the species best, guiding them in where to designate set-asides from harvest. "What made it possible was not just a law but a president willing to use his power," Franklin said. "One of the other keys was a crisis." With logging shut down by the judge, action was necessary. And with the scope entirely on federal lands, the problem was squarely within the purview of the federal government to resolve. The Northwest Forest Plan took sweeping effect in 1994, on management of more than 24 million acres of federally managed lands in Oregon, Washington, and northwestern California. The reserved lands included 79 percent of the acreage across seventeen national forests, 11 percent of the forestland in seven Bureau of Land Management (BLM) districts, and 9 percent of the acres in six national parks. In all, 30 percent of US Forest Service and BLM lands, or 7.4 million acres in the range of the spotted owl, were put off-limits to cutting, protecting trees eighty years and older in so-called Late Successional Reserves. The plan also designated 10 percent of federal forestland, or 2.6 million acres, for protection of lands next to lakes, streams, marshes, and wetlands. Another 4 million acres were designated matrix lands for multiple uses, including logging on sixty- to one-hundred-year rotations.

The Northwest Forest Plan was the largest shift in management focus for the US Forest Service since it was founded in 1905, from timber harvest to conserving biodiversity, with an emphasis on preserving endangered species. The litigation that led to the plan (and the plan itself) was never really about just the owl. The bird stood for the integrity of an entire forest landscape and its ecosystems on which its survival depended. "What I am extremely grateful for is the spotted owl doing the job of being the flagship for the old-growth forest," Franklin said.

> Without the spotted owl, it would have been very difficult for things to develop the way they developed. This wasn't really about the spotted owl, but about old-growth forests. We can honor the owl for what it did do, which was to focus us on and epitomize the older forests. And I have a lot of gratitude to the northern spotted owl. It really wasn't about the owl, it was about a system, it was about forests; what we learned was to open our eyes to what a forest is really about, as opposed to a plantation. The owl was a doorway to an education about the old-growth ecosystem, it was such a symbol of it. It accelerated our learning, and it resulted in a widespread audience having gained knowledge of what the forest was all about. It was a wonderful confluence, and it made things possible and facilitated events that might have taken much longer. It is just really wonderful when one of these kinds of animals is such a symbol for a system.

The Northwest Forest Plan itself changed everything but resolved little, Franklin said. It was not protective enough, allowing logging on mature natural forests he would come to regret leaving unprotected. As cutting cratered on federal lands from about five billion board feet a year to one billion, the clear-cutting continued unabated on private industrial lands. Putting a stop to clear-cutting on federal lands also didn't begin to curb the effects of other threats to species and habitat—from cutting elsewhere to climate change, high-severity fires, habitat fragmentation, toxics, altered streamflows, shifts in ocean conditions, and invasive species. The northern spotted owl, marbled murrelet, and Pacific salmon all have continued to decline. Perhaps not as rapidly as they would otherwise, but listing for protection under the federal Endangered Species Act and habitat set-asides under the Northwest Forest Plan were not the rescues that were hoped for. Fires have roasted eastern Cascade old forests. Most devastating has been the invasion of the barred owl in the Pacific Northwest since about the 1970s. The arrival of a wide-eyed, fluffy, less than two-pound bird has turned the food web and ecological integrity of these forests, protected at such effort and social and political disruption, on their ear. For this is an animal that will eat almost anything, and voraciously.

"Habitat doesn't mean a damn thing anymore," Franklin said. "We

This forest was clear-cut and converted to a plantation—a farm for growing wood as quickly and efficiently as possible.

preserved millions of acres of habitat and it hasn't done a thing to prevent the catastrophic decline of the [spotted] owl populations. The tradition in conservation biology is if you just preserve habitat, it will all be okay. But not when you have an invasive species like this [the barred owl]. This isn't just about the spotted owl, it is about a range of organisms facing a top predator they have never faced before. It eats other birds, bats, crustaceans, amphibians; these organisms have never had this intense predation because it is omnivorous. The barred owl population has become very dense—you can cram one helluva bunch of barred owls into a space which a spotted owl would have occupied. We have a new top predator." Franklin ruefully recalled taking a group of students on the Asahel Curtis Nature Trail in North Bend, Washington, and encountering a glorious Pacific Northwest giant salamander. "It was just sitting on a stump, and I thought, 'You are dead meat for a barred owl.' And it was just marvelous. It was cold, so not moving much, and it was about a foot long. All I could imagine was, *You poor critter, you don't understand in this forest, you have a predator that will take you out.* The barred owls are like vacuum cleaners. If it moves, and they can catch it, they will eat it." In the summer of 2024 the US Fish and Wildlife Service decided to go forward with a controversial kill plan deploying trained shooters to take out as many as 450,000 barred owls in three states over the next thirty years to make space for northern spotted owls to persist. Managers say they have no choice but to cull one owl to try to save another that is rapidly headed to extinction.

Not only was the Northwest Forest Plan not enough to save the northern spotted owl, it was also a geographically and ownership-limited rescue only for old-growth forests on federal land within the owl's range. Just across the US–Canada border in British Columbia, old-growth forests continued to be commercially logged, as they are today, even with all that has been learned about the unique and irreplaceable value of these forests. These are the same forests that nurture the same biodiversity and provide the same values for climate protection and human well-being. Yet just across the Strait of Juan de Fuca, in the same bioregion, these forests continue to fall, simply because they are across the border. Only about 3 percent of the biggest old trees outside of protected areas are left in British Columbia, and every year the cutting continues. For some, the losses feel personal. Because they are.

4

SALMON FORESTS

BIOLOGIST TERESA RYAN has seen a lot of clear-cuts and loaded logging trucks. But this trucker was hauling logs down the mountain outside the remote logging outpost of Woss, British Columbia, cut from trees so large the tractor trailer could carry only a few. "It's so sad, I just feel heartbreak, like a piece of me just went down the mountain," said Ryan, whose traditional name is Sm'hayetsk, as the logging truck roared past her. "The ancestors are there. We actually used to put our people in the trees," she said, speaking of traditional burial practices of the Gitlan of the Tsimshian Nation. Ryan explained that her ancestors put the remains of their people in bentwood boxes and hoisted them into trees, into the life of the forest. So when she said the ancestors were in those trees, just cut from the mountain where they had lived for centuries, she meant it literally.

Ryan is an Indigenous Knowledge and Natural Science Lecturer in the Department of Forest & Conservation Sciences at the University of British Columbia (UBC), Vancouver. I had joined her and Suzanne Simard, an eminent forest ecologist and professor in the same department at UBC, in these woods for a week of fieldwork. Ryan and Simard had gone out on this logging road to find their crew. But right now, that was out the window. They had to go see the clear-cut those logs had come from. Ryan jumped in Simard's battered field truck, and I squeezed into the back seat with their clipboards, batteries, and gear for sampling and measuring trees. Simard bucked the truck up steep, rugged slopes to reach the top of the cut and killed the engine. We got out and looked around at the cutover land, bare and baking in the July sun. A grapple skidder awaited the next day's haul of logs, and sawdust lay fresh over the ground. "This kind of machinery is devastating to the forest floor," Simard said, eyeing the carved-up ground, rutted and scraped to mineral soil. "And we are left with that. There is no carbon left, it's gone, and it's never coming back. It took twelve thousand years to make this, and we have lost it in a snap of a finger." The air smelled sharply of ground-up fir needles.

"You can't stop it, these trees are already down," Ryan said, assessing the stumps, slash, and logging debris. "It was coming straight at me," she said of the logging truck. "All I could see was big logs. They were so huge." This cut was not exceptional; as we walked to the top of the cutover hilltop and took in the landscape around us, we saw logging roads tracing through a maimed landscape into the far distance, mangy with clear-cuts so large they covered entire mountainsides. "It's a tree hearse," Simard said of the logging truck, as we walked back to her truck. "We are down to the last drops. It's a fucking graveyard out here. These are some big trees. They *were*," she said, correcting herself. "There is nothing we can do. I know we can't save every tree. But we have to do a better job. We are going to need wood, but it should not be the old-growth. That was all cedar. It just makes me feel sick. And the fact we have seen it, day after day after day, all over the country, I'm sure those trees are a thousand years old. People grow so numb to it, we all are."

Ryan gathered a scarred piece of wood from the ground as we walked back to the truck. "We've got to burn this, to honor the spirits. Honor the ancestors," she said. As we headed down the mountain, a distant rumble filled the air. "Here comes another," Simard said, yanking the wheel to pull over to get out of the way of the log truck. After it passed, as we descended the mountain, ravens were circling and calling. "Stop," Ryan said. "Stop right now." Simard jerked the truck to a halt, and the ravens' calls filled the still air. "They are crying, their home is being taken," said Ryan, who is of the Raven Clan. "Their nest trees are being destroyed. That is the mom. That is the baby. She's in shock." We just sat for a minute, quiet. Then Simard

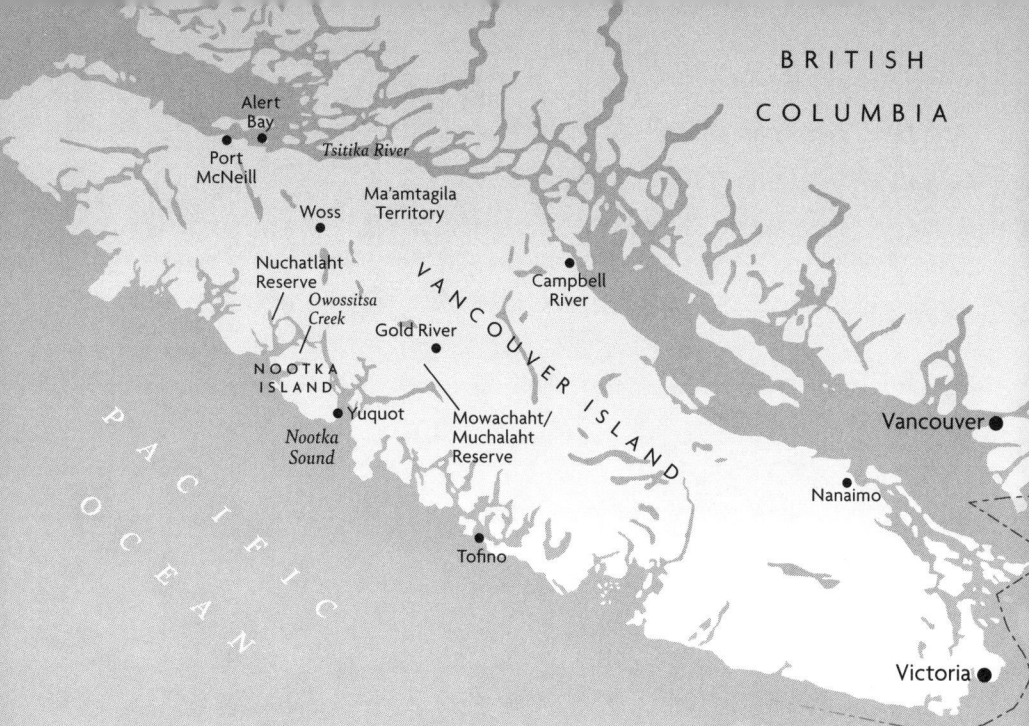

Vancouver Island

started back down the logging road, jouncing over the ruts. "It's so disrespectful to the mountain, it puts the mountain to shame. Our shame," Simard said, gathering speed past a numbered sign for the logging road. It was riddled with bullet holes. We made the long, bouncing drive down to the valley bottom, where we hoped to rendezvous with their crew, in silence.

Called the Mother Tree crew, this was a team of researchers Simard had assembled in 2015 for a research project creating a time sequence of the amount of carbon stored in these clear-cuts—in the first pass, the second pass, every time the loggers came back. They wanted to learn, through analysis of the soil and its layers, the toll taken by each cut, compared to the original soil baseline condition. "It shifts down, and down and down, they will come back and plant it, and it will come back to that pale green, impoverished condition, a second-growth stand, with no variety in the canopy," Simard explained. It

Old-growth is still being logged in British Columbia. This tree was felled on Vancouver Island, already a sea of clear-cuts.

was soil that started her career, and now soil that had made her internationally famous, as the author of the bestseller *Finding the Mother Tree*, which tells the story not only of Simard's scientific work but also her struggle to overcome skepticism of her findings. Simard's lab explores the role of mycorrhizal fungi in connecting trees, one to the other, sharing nutrients and communication, all around hubs of the largest trees she dubs "Mother Trees." When her work was published in the journal *Nature* and popularized by the press as the "Wood Wide Web," likening the networks to the internet, Simard was suddenly an internationally famous scientist. Simard, who had labored in obscurity and hostility from industry and even her own colleagues, was a sensation, with millions of views on her TED Talks and a crush of media attention. She bore the attention stoically, doing interview after interview for the sake of the forest—which, even as her notoriety grew, was still being cut down.

Protests during the summer of 2021 against one logging operation, near Fairy Creek at the south end of Vancouver Island, sparked the largest civil disobedience in modern Canadian history, with more than a thousand arrests of protestors trying to blockade the logging roads. They dug trenches in the dirt logging roads cut into the sides of remote mountains. Locked themselves to concrete blocks, and erected platforms in the middle of logging roads they then locked themselves to. They camped in the canopy of towering firs. They hiked miles to the leading edge of the cut, where the loggers were dropped off by helicopter for their daily work—amid protesters leaping from the understory blowing air horns. They wrote peace signs on the cut stumps with fresh sawdust, scooped from scarred ground next to gas cans left by the loggers for the next day's cutting. The Royal Canadian Mounted Police, backed by the courts, hauled away the protestors. The logging continued—but the provincial government of British Columbia was beginning to talk about deferring logging in some areas, with the effect of spurring even more logging. "They are cutting it faster and faster," Ryan said. "They've got to get it quick before it's locked up forever, they are cutting hill after hill. It's apocalyptic."

Simard bumped over more logging roads until we saw the crew's work trucks. We parked and headed into the forest. We hopped from piece to piece of downed logs over inky-black mud in a fructifying bog at the edge of the road, following a faint trace of the crew's path. In here, the light was soft, the ground even softer, and it was cool, even on this blazing July day. Fallen logs were crumbling back into the earth, where they started as seedlings centuries ago. The logs were soft to the touch and furry with moss. Lush lichen grew over them, leafy as lettuce. The sunlight diffused through needles of hemlock and fir, and long shafts of gold, late-afternoon summer sunlight found canopy gaps and gilded the forest floor. Far above, a breeze was stirring green branches across the blue of the sky. Simard had found her research crew, measuring giants in this old-growth stand. It was day's end and time to wrap up. When they had gathered their equipment, we headed back to camp.

Mak'wala (the traditional name of Rande Cook) gathered us nightly in a circle after a camp dinner to reflect on the work underway. He is an artist and hereditary chief for the Ma'amtagila, one of the eighteen tribes of the Kwakwaka'wakw whose territory reaches from northern Vancouver Island southeast to the middle of the island and includes smaller islands and inlets of Smith Sound, Queen Charlotte Strait, and Johnstone Strait. In 2022, Cook had invited Simard and other scientists, artists, filmmakers, Indigenous knowledge keepers, and writers including myself to the second annual Tree of Life Project, for a weeklong exploration of the old-growth forests in Kwakwa-ka'wakw territories. Founded by Cook as the Awi'nakola Tree of Life Foundation, he partnered with Simard from the start. He saw parallels between his people's understanding of the connections between all living things and her work focusing on healing damaged forests. The goal on this trip was to share both new and ancient ways of under-standing the land, and to directly experience together the impact of industrial logging on the land, water, and Native cultures. "Right now this is really about the next one hundred years," Cook said in camp that first night. "That one hundred years doesn't belong to us. In a very short amount of time, the amount of damage we are doing is irreversible." It was time, he said, for a societal shift from devastation to regeneration.

Simard's goals were in alignment with Cook's: to use Indigenous, scientific, and artistic knowledge to understand and heal the forest. In her Mother Tree Project based at UBC, Simard and her team were assessing the biological and ecological status of clear-cuts, comparing their soils with what they found in intact old-growth forests, to get a sense of a baseline condition. They were like arboreal ambulance chasers, trying to stay one step ahead of the logging crews, searching out the last of the unprotected old-growth in these lush river valley bottoms, hurrying to learn what they could of what intact old-growth forests look like and how they function before they are gone. "We are trying to find old-growth in these ravaged landscapes; all we find are rare patches, scraps perched on cliffs," Simard said as we drove to the day's research site—the places the day-trippers and vacationers to Vancouver Island rarely see. "We will never know what we have lost. We are at a point in history where we can never understand what we once had. But there are still clues—what seeds are buried in the ground, what soils—that is what we are doing, trying to reconstruct what is there, so we can heal these landscapes."

This is the Vancouver Island off the main highway, with its beauty strip hiding the views of the cutover slopes and valleys. We were in a hub of logging roads, logging signs, clear-cuts, loading areas, and equipment yards. This land is hardly a forest at all anymore; it is a manufacturing plant for timber products, with remnant old-growth stands amid a sea of industrial tree farms on their third cutting. It is the nature of her work that Simard has to drive through active logging areas to access her research plots. As we drove on, dust boiling behind us, the caravan of research trucks and the Tree of Life crew ground to a halt: there was a chain across the road. "ACTIVE LOGGING AREA" read the sign on the chain. We smelled it before we saw it: the unmistakable odor of ground-up freshly cut trees. "Those are old-growth trees," Simard said, glancing at logs piled by the side of the road, on the other side of the chain. She swung out of the truck to talk to the logging crew, to explain she needed them to let her and the rest of the crew pass. After a brief conversation Simard returned, dropped the cable to the ground, and the procession of vehicles drove through. We regrouped by the piles of logs heaped by the road to be taken away for milling. The cuts were fresh, weeping sap. The bark was still fragrant, the wood moist to the touch.

"I'm just numb," said Cook. Logged on his people's unceded territory—a five-hundred-thousand-acre swath of Vancouver Island. This displacement of his people from their lands, of the trees from their land, of the wildlife from their forest, continues the legacy of settler colonialism, Cook said. "Nothing has changed since the beginning," he said. "Those policies are designed for these actions to happen, and for us to say something, we are the criminals, we are the disrupters. We are so conditioned by society to say, 'This is okay, we support this. I need an extension on my nice home. Why do you guys get in the way, why are you so disruptive?'" Cook said he formed the Awi'nakola Tree of Life Foundation out of desperation and as a matter of cultural survival. "We are watching our own traditional territory be wiped out, demolished; we are down to the last 2.7 percent of old-growth, we have got to get it all, it's sick to me. I think more and more it's like a panic, that we are getting closer and closer to having nothing." But he wonders, "Who are we without our forests? How can we carry on if we don't have this connection to this land that's at my absolute core. It's severing our tie. Culture is not a performance, it's connection. That is how it is in our culture. We can't just sing and live disconnected in order for there to be new songs, for it to be a living culture."

We kept going, to get into the uncut forest, to lay out some sample plots and dig. Simard led the way, walking right over the ridge on the opposite side of the road from the heaped old-growth logs into the trackless deep forest beyond. The steepness of the slope to the valley floor did not slow her. She was headed with her sampling crew to a place the maps showed should have what she called "the white rhino of the forest": enormous, untouched old-growth forest, reigning supreme for centuries deep in the heart of the Tsitika valley. Unprotected yet still uncut. We made our way across downed logs over streams, through bogs, the islands of forest floor in between the muck carpeted with sphagnum moss. The coolness coming off the Tsitika River reached us before we found the trees. Even on this hot mid-July day, it was so cool in the shade of giant hemlocks and Douglas firs we kept our coats on. But the sun found us too, aglow through gaps in the canopy where bigleaf and vine maple surged into the light. A goldfinch spangled in the sun, a brilliant yellow flash amid the river's sparkling blue.

RYAN AND I HEADED TO THE BANKS of the river to bask, and to talk about her work in Simard's lab, tracing the nutrients of salmon in the forest, their nitrogen a vital boost for these old trees. Nitrogen in soils enriched by decomposing carcasses of spawned-out salmon has a unique chemical signature that can be traced in the trees that draw up that nutrition. This signature of salmon in the trees is one of the characteristics of these old-growth forests. "Forests and salmon are connected," Ryan said. "And they always have been. Without the forest we wouldn't have salmon. And we wouldn't have these forests without salmon. The trees are part of these cycles, all the way to the mountaintops." Shrinking the size and reducing the abundance of salmon returning to these river landscapes thus breaks the cycle of nutrients coming home to the land from the sea, Ryan explained. This starves the forest as well as the animals living in it.

Cutting the big old trees is part of what has diminished, threatened, and even endangered salmon runs all over British Columbia. Pacific salmon are creatures of the forest. They are born in woodland streams and return to their freshwater redoubts, even swimming high into mountain watersheds, after two to seven years foraging in the great pastures of the sea. They return home to spawn in the gravels of cold, clean rivers that thread these ancient forests. The biggest fish, including Chinook, use the main channel, while others, such as coho, make for quiet side-channel hideaways, rich with food and warmer in winter. The forest's dead and downed logs provide the complexity and structure in these streams, bending the flow to the quiet side channels where tiny salmon fry rear and feed. The shadowy pools dammed by big logs and tangles and jams of big wood in streams all provide hiding, resting, and feeding places for salmon at every life

Teresa Ryan (Sm'hayetsk) of the Gitlan tribe of the Tsimshian Nation brings both her scientific training and her ancestral knowledge to the study of old-growth forest ecosystems, especially the relationships between salmon and forests.

stage. Downed logs stabilize the streams at high flows and the big trees hold the banks, where the intimacy of nutrient exchange happens. Needlecast, leaf litter, and a rain of insects from the forest all feed the streams—from fish snapping up juicy insects to the tiny creatures of the benthos, the grazers and shredders chunking the forest litter down to size. It all counts, from the algae on the rocks to the caddis fly larvae trundling on the bottom in their spun casing, waiting to hatch.

In the same way that a live and dead tree are really just a continuum of the same being, the forest and its streams and the beings in them are one interrelated organism and force of life. It is this complexity and interrelatedness that makes these dynamics inseparable, and the wealth of each dependent on the other. "What we see are these cycles and connections, and when the connection is broken, the cycle stops," Ryan said. "And what we see is these cycles are breaking down. There are changes in the fish, changes in their size, changes in their run timing. And some of these things can't be seen by science. We have a baseline within our Indigenous knowledge system embedded in our stories, our cultural practices. This is what is meant by 'our way of life.' When all these connections are moving in harmony, we have enormous benefit." When they are not, she explained, people and forests and rivers are impoverished. The Pacific salmon come home following their sense of smell to the very river where they were hatched, traveling thousands of miles from the open ocean. Ryan wonders if the fungal signature of soil in the groundwater from the hyporheic zone (which steeps the gravels where the salmon make their nests) could be part of how they find their way home with such precision. "It's the synchronicity of these events that occur with the salmon cycle, something about how they select their site," she explains. "Chinook need groundwater to be consistently moving through the gravel, to keep it at a constant temperature, the hyporheic and the groundwater provides the oxygen and temperature, and she is going to pick those sites. I believe the connections belowground, the tree's root system and the fungus that generates an olfactory sense, is what makes this their home."

The salmon return the favor to their home waters. Their spawned-out carcasses feed the waters in the hyporheic, interlaced with the roots of the trees. The bears, the wolves, and other creatures that eat salmon drag those carcasses into the trees, and their excrement carries the nutrition even farther into the forest. Salmon feed the land many miles from the sea, and Ryan is interested in whether these nutrients perhaps are pulsed more thoroughly and expansively inland than has been previously understood, through the mycorrhizal fungal network between trees. It's all part of a beautiful cycle she is seeking to document, between animals, fish, the rivers, and these old trees. When the salmon are abundant, predators take just their favorite part, usually the brain or the belly, and leave the carcass in the forest for some other scavenger. And the flesh also goes back into the soil. The bones and fins stay at the surface longer: calcium that dissolves into the soil and into the mycorrhizal networks connecting the trees' roots. In this way abundant Pacific salmon are part of the food security of wildlife and these salmon forests, as well as that of the people.

Salmon gain 95 percent of their body weight at sea and provide a major source of carbon, nitrogen, and phosphorous for their own young as well as for entire ecosystems of the Pacific Northwest when they return to their home watersheds, gifting those nutrients to the land and the waters. For this reason salmon are a so-called keystone species, supporting overall ecosystem health and tying entire watersheds together. A study of this web of life was first published by state and federal scientists in Washington and Oregon in 1990. The study was expanded in its scope and contributors to include timber companies, nonprofits, tribes, and hydropower managers. The result of their work, *Pacific Salmon and Wildlife: Ecological Contexts, Relationships, and Implications for Management*, was reissued in a special edition technical report in 2001, documenting the links between salmon and wildlife species as well as the broader aquatic and terrestrial realms in which they coexist.

Revealed in the report was a beautiful codependency between salmon and the animals that rely on them and the reliance of salmon on the animals. Bears, bald eagles, river otters, and more transport the salmon carcasses that feed the water and the land. Beavers build dams that create side channels where salmon rest and feed. Nearly 150 wildlife species depend directly or indirectly on salmon in their various life stages, the researchers found—from killer whales and river otters and giant salamanders to grebes, turtles, and aquatic garter snakes. Insects such as stone flies, caddis flies, and blackflies feed off salmon carcasses and become food themselves for juvenile salmon. Improving the health and abundance of salmon feeds the rivers, the land, the forests, and the animals that rely on them. Salmon are the silvery shuttle that interweave the region, with their migration from interior rivers through the estuaries to the ocean and all the way back again, bearing gifts from the sea to the rivers and the land.

People too are woven by salmon into this web of life. Native cultures have relied on and been enriched by Pacific salmon since the end of the last glaciation and the beginning of their settlements in the Northwest on both sides of the US–Canada border. "We have this connection, we are talking about a food source that for thousands of years supported communities of healthy people," Ryan said. "The chronic diseases, the diabetes and digestive issues, we never had those. Our diet has changed." Ryan is the first in her family to go to college, charged by her grandfather with getting a PhD, to bring her Native knowledge to bear with science to improve the health of these forests and streams. It was a very specific charge she dared not shirk, Ryan told me, despite its hardships. She was so poor as a student, when she was doing her undergraduate work at Central Washington University, she would make a small cedar basket, put it on a string of leather lacing, and visit the staff in the office. "The basket would be admired, and I would get asked where I had gotten it," she explained, "which led to the question of selling it, and then I would have enough money to buy some milk."

Today Ryan works at the intersection of ancestral knowledge and science in biological and cultural systems as complex and beautifully interwoven as the baskets she sold to put herself through school. She works with the Mother Tree Project to bring Indigenous knowledge and the natural sciences to the center of the lab's work on the future of forest management, exploring how old-growth forests support biodiversity. "The reason I do what I do is to be sure there is a future in these forests, in the salmon, and in our people," she said. "The future is our management system. It is not about this year's catch." Wealth in her culture was always counted in what would be there for future generations in their home territory—never in money to be made for individual profit, stripping a place, and moving on to the next one. As a scientist, Ryan is appalled to see non-Indian fisheries professionals estimating reduced populations with ever greater certainty, with fish counted like coins for the taking. Her teachings instead regard fisheries management as a matter of stewardship, with balance and reciprocity animating the question of how to be in relation with a community, including salmon, the herring, the forests, and all other nonhuman beings.

In this way of thinking, the world gets much bigger, with connections between species acknowledged, interactions respected as needs that matter beyond our own, and responsibilities to the future awarded as much importance as rights to harvest in the present. This thinking acknowledges and protects connections between communities, human and non, and between the past, present, and future. "This is a fundamental component of who we are, and our connection with the lands and waters," Ryan said. "There is no separation between the two. Without cedar there's no water. There's no salmon habitat. These things are so connected." Aligning her Indigenous knowledge systems with her scientific research has taken her thinking about salmon beyond data. "We are counting them to extinction," she said, while not significantly altering the forestry or fisheries practices that are ensuring the decline of both forest ecosystems and the salmon that depend on them. This too was the message of the technical report on salmon and wildlife, calling for management that looked at salmon as more than a commodity for human harvest, and instead sees and treats them as the nourishment for all beings that they are. Taking that seriously would have implications for salmon management, including fisheries practices, the report found.

Ryan is investigating the role traditional management practices can play in rebuilding the size and abundance of Pacific salmon. She has been piloting a bring back of traditional Indigenous stone salmon traps, designed to allow passage of the biggest females into the streams, to restart the genetic inheritance of bigness in future generations of fish in runs that have been mined for their largest fish for a century. The traps are inundated at high tide, allowing fishers to take the fish they want and let the rest go. The result is a limited harvest and one that leaves the biggest fish for the forest—and the future. "You can walk in and pick up what you want, and leave the big ones. That is our Indigenous knowledge and our practice," Ryan said. It's a fishery that depends on knowing the cycles of the salmon, and the tides and the phases of moon, as well as understanding the variabilities of flows and migration timing, from stream to stream. The chiefs traditionally were charged with taking care of the fish, which included knowing this and other ecological truths, such as the importance of woody debris left in streams for the adults to take cover. These stewardship practices and teachings were to result not in money in a distant bank but in community wealth, in an abundance of healthy food celebrated and shared at a feast. That was the ultimate test: abundance.

Having enough food to feed everyone and invite guests to a feast. The more food a people could give away, clearly the wealthier they were. *That* was high-class status. "The feast, the potlatch was where you could see if it was working," Ryan said. "The potlatch is the agency for the power of our law and our practices." This was also wealth for trade, for putting by for the winter season, and wealth returned to the streams in living salmon that laid their eggs for the next generation and fed the stream and the land with their bodies. This is the original diversified economy, with something to eat in every season. Salmon and the beings they feed—from trees to berries to bears, wolves, birds, medicinal plants, and more—were a perpetual abundance if cared for correctly. Salmon were a silver shine of wealth coming back to the forest and streams and the people, fresh from the sea, not only for the expediency of now but for the future. Central to that wealth is the duty to take care of it. "We are reinvigorating what we have known for thousands of years," Ryan said. This is a different approach to the Western science understanding of the salmon cycle, of thinking of it narrowly as shuttling of salmon from egg to ocean and back again, to set fishing seasons for people. Where is the allocation of salmon for wolves and bears and orca? For the soils, for the forest? For generations of people of every sort yet unborn, who will rely on abundant thriving forests and streams?

"I have just explained a completely different salmon cycle, as a way of life, connected to people," Ryan said, as the river rushed by us. Just as important is a different understanding of wealth and success that elevates respect and relationality. The First Salmon ceremony is widely practiced in Coast Salish cultures. Each community has its own version, but in general the first fish caught of the year is given to elders, and the bones of the fish returned to the stream where it was caught. This bears great respect from the people for the salmon chief in their village under the sea, so the salmon will come back. This ceremonial practice is not incidentally a return of calcium from the fish back to the system that produced it, another sharing of wealth between the salmon people and tree people and human beings. "These are all beings," Ryan said. "My being is no different from this. These are living beings and we know when we use these things, they are living. It's understanding that we are part of this whole cycle, and it's connected to our food security, it's medicine for the elders, to the young people, so they grow strong and carry on."

These links between ecological and social systems and cultural teachings are intimate and of great antiquity; it is only in the past 150 years that they have been ruptured by settler colonial, capitalist, extractive systems, and displacement of the caretakers and their knowledge systems from their place and their sources of wealth. To be successful, Ryan said, ecological restoration must bring together scientific understanding of how forest and river ecosystems work with these knowledge systems and practices, informed by thousands of years of sustaining abundance. There can be no ecological recovery without cultural recovery.

RYAN AND I HEARD SHOUTS of excitement coming from the forest. Turning away from the river, we headed up to where the Mother Tree Project crew was beginning to measure the height of old-growth trees in this grove. "Seventy-four meters! I've never measured a tree that big, holy cow!" said Liam Jones, a six-year veteran with the crew. He gave the 242-foot-tall Douglas fir a kiss and a pat. "We never get to work in forests like this, not at all," said Jones, as he laid out the research plot to inventory the trees and other species at the site. "These are the biggest Douglas fir trees we have ever worked with, it's hard to put into words. It's incredible, so overwhelming and emotional and exciting. Especially after this morning, seeing all those massive trees piled up by the side of the road and lying there. To walk into this, now, it's almost a relief. It's so calming to be here, it just resonates."

Author and scientist Suzanne Simard with her crew from the Mother Tree Project, studying an intact old-growth valley bottom forest on Vancouver Island. This vanishingly rare and unprotected forest, she predicted, would soon also be cut.

Zoe Neudorf, a field assistant with the Mother Tree Project, shook her head and laughed with delight at the inability of their instruments to calculate the tree's height; it was impossible to get an angle on so tall a tree. The team did not have a borer even close to big enough to take a core, to assess the old tree's age. Everything was on a grand scale: the understory was lush with moss and sword ferns. Delicate foamflower with its white blossoms glowed in sun flecks, and the tree trunks were eared with conchs and green with lichen. No wonder the team was in awe, after the industrial plantations they usually work in—dark, barren, the light choked by the dense planting of an even-age monoculture. This was truly a *forest*, alive with diversity of species, size, light, and soils. Its root system was intact and connected at every layer, never compacted, scraped, ripped, burned, or sprayed with herbicide.

I became aware, amid the hoots and hollers of the crew, that Simard was working, off by herself, digging a soil sampling pit amid the old-growth giants. Her favorite place to be and thing to do. Simard has built her career on soil, probing the wilderness belowground, to explore how forests work. Through her Mother Tree Project she and her team are building the scientific basis for a rescue plan for forests that have been cutover and replanted to monocultures, to investigate ways to heal these damaged lands. Are there renewal practices that can protect biodiversity, carbon storage, and forest regeneration, particularly as climate changes, putting even more stress on forests? "It's not just how do you protect these forests, it's how do you restore them," Simard said. "This is a landscape that's heavily trashed. And there are thousands and thousands of species we have never even described. How do we restore these forests that took twelve thousand years to build? It's working with a damaged landscape. How do we get that oppressive monoculture canopy off, bring back the salmonberry, the thimbleberry, the seeds hidden in the forest. It's the understory, how do we open these forests up, release that vitality? We need hand fallers. We need to demechanize, deconstruct, to reconstruct." Descended from a family of hand loggers in rural British Columbia, Simard wants to get people back on these lands to heal them. "This is an urgent thing. There is a shift happening. There is a reason why we are all here. We can work with these forests with our hands and our hearts."

Within the square-meter microplot she had laid out in the old-growth grove, she began a soil sample, one of ten that would be done in this area, to gauge the amount and type of biomass and carbon stock in a hectare, or 2.5 acres of this forest. Her tools could not have been simpler: a field shovel, paper bags, pencils, a clipboard, a ruler. By the end of this week in the field, one of the motel rooms reserved by the team would be piled with hundreds of bags of soil going back to the lab. There they would be oven-dried and tested for carbon and nitrogen content. Simard had the river's voice for company and she settled into the rhythm of the work, separating samples of the shrub layer, herb layer, moss layer, forest floor, and fine woody debris in her plot into bags. Although it looked simple, this was actually work probing a new frontier, looking more deeply into the soil than conventional sampling to discern its makeup, across a range of gradients of elevation and forest types. The intact coastal rain forest stands—what few she could find to sample—would, she knew, be off the chart for carbon content, in the same league as tropical rain forests in the Amazon, she predicted.

Using her fingers Simard pulled off first the duff, a fluffy layer of needles, leaf litter, and fern bits. "This is the thing everything wants, its carbohydrates, and they start eating this stuff, the shredders, the rippers, the grinders. They make it smaller and smaller and smaller," she said. "Those are eaten by bacteria or fungi, and they are attacked by nematodes and mites. They are eaten by tardigrades and spring-tails, it becomes more humidified, all that poop and pee, the plants can take that up as their nutrient cycle works at this top layer," Simard said, pulling the duff layer apart with her fingers, showing me its rich feast. Then she dug deeper into the top soil, moist and dark, the product of centuries of decomposition in this forest. Still protected by the shade of an intact canopy, this soil had retained all of its moisture and was soft, spongy, rich, and fluffy. "This is the gold," Simard said, oblivious to the clouds of mosquitos working on her just as assiduously as she worked on gathering her samples.

It's a job she seeks whenever she is out working with the crew. "You have to be able to slow down, enjoy the quiet moments, it's a simple job. There is no job that is too humble," Simard said. "What did you do this summer? I picked up sticks and counted carbon content on the forest understory." She said this with pleasure, teasing apart tree rootlets with her fingers, revealing the mycorrhizal fungi at the root tips. "This is a really, really rich site," she said. These fungi grow into tree roots, to take from them carbohydrates the fungi, as an organism living belowground, cannot get from sunlight. But the relationship does not stop there—the fungi does not only take, it gives. By its presence in the soil it vastly expands the surface area of the tree's roots, gathering water and minerals it passes to the tree, even as the tree's succor passes to the fungi. Simard calls these fungal networks, and the largest tree hubs in the network, Mother Trees. Her lab has mapped them in various forests around British Columbia, and she and other scientists are exploring how they work. So far, she has found the hub trees pass nutrients in greater amounts to genetic kin—hence her naming them Mother Trees. The work has documented as well that when these hub trees are removed, such as with logging, the fungal network is disrupted, broken, and removed, depleting the forest's regenerative capacity.

These are the biological legacies crucial to the forest's reset after disturbance, ensuring continuity to the next generation—just as Jerry Franklin had found in the old-growth forests he studied in the Pacific Northwest of the United States. And like Franklin, Simard's curiosity about how forests work also began with wondering what we really know about old-growth forests the agency she worked for was bent on cutting down. Her first job was as a forester, delineating the boundaries of forests to be cut, including old-growth. What struck Simard was how often the cut areas subsequently treated with herbicide and planted to a monoculture for the next harvest did not thrive. She wondered if the loss of companion understory plants and their suppression with herbicide, the taking of all the trees in a stand, and destruction of the soil by heavy equipment and replanting with a monoculture were themselves restricting the very thing foresters wanted: quick and thick regrowth.

Simard's research required working in bear country in backcountry British Columbia, with primitive equipment, sometimes with her own family helping lug gear to remote sites, including dangerous radioactive isotopes used to trace the movement of photosynthate—the food made by trees through photosynthesis—between one tree seedling and another. In her early work she placed a bag over one tree so it could not photosynthesize and inoculated another tree next to it with the isotope to be able to see what, if anything, was shared between them. And sure enough: the photosynthate from the tree with its head in the sun could be tracked in the hooded tree's tissues, moving through the mycorrhizal fungal network between them. This was revolutionary in what it implied for forestry practices. It showed that trees exist not only in competition but in community. It followed logically that by clear-cutting, eliminating the understory and leaving the fungal network depleted, tree regrowth would be set back—and sometimes even defeated.

Simard's findings were (and still are to some) controversial because they challenge the efficient, industrial processes used in commercial tree harvest for decades and point toward the need for retention of the biological legacies that preserve the resilience of forests after a disturbance, whether natural or human. That costs money but in a time of climate change, with more stress on forests than ever, Simard is convinced practices have to change. Her Mother Tree Project is a collaboration across forestry disciplines to investigate how forests fare if they are allowed to be more like a forest, even when cut. Franklin was onto a similar thing on his side of the US-Canada border, taught by his work at Mount St. Helens about the importance of biological legacies in the restart of life after a disturbance. His ecological forest management practices—which discourage most if not all herbicide use, much greater retention of trees left uncut during harvest, and replanting of a diversity of species including understory plants—have been adopted by some public lands forest managers. But not by industrial landowners managing their lands for quick and maximum financial return. Managing primarily for one value will not produce a balanced result, I often heard Franklin ruefully say in our walks and talks both in cutover lands and the real forests he so prizes in the Pacific Northwest.

Simard was reaching deeper into her soil pit, using her bare hands to feel the earth's texture. "I go down until I hit the mineral layer, I feel the grit," she said. This soil profile she was encountering here was totally different, she said, than in a managed plot that had been cut and planted at tree densities far beyond natural conditions, with more than a thousand trees per acre, instead of about four hundred. "It's really dark, there is no understory, no salmonberry, no huckleberry, no ferns, just a lot of trees. Herbicide spraying ensures no understory will take root in the clear-cut while it is still open to the light. Sampling the soils there is a different experience." Disturbingly, she explained, "there is no sound of roots breaking. There is nothing there, it is very visceral. When there are no roots, the soil collapses, there are no pores, it is like cement, and it is highly erodible, the roots are gone. Soil structure is so important, when you lose the soil structure, you lose the fertility. And these roots are always excreting stuff, exudates and mucilage, that is what holds the particles together and makes it crumble, that aerates the pores in between. All the creatures function better than in more gloopy stuff."

Simard checked the volume of her sample pit, using water in a plastic bag. Satisfied she had done the sample properly, she stood and dusted herself off. "It all comes from this humble, sampling place, from the ground up, there are so many things to look at, measure, to restore. That's the joy of rebuilding our world, then we feel like we are doing something instead of just being in despair." Surrounded by paper bags she had just filled with layers of the forest, she said, "I used to eat dirt. I got a lot of worms. My mum just put me in some dirt pile and I'd eat dirt. My brother and I would have lump fights. It wasn't until later I discovered, 'Oh my God, there is this whole field called soil science.'" Simard's specialty has both fit and shaped her personality: "When you meet soil scientists, they are very humble people. They deal with dirt."

Laughter found us, from where the crew was still working to inventory the trees in their sample plot. "This is not big enough, it's only a five meter dbh [diameter at breast height] tape," Jones, the field tech, called out, delighted to find a tree too big even to measure with the equipment they brought, sized for the second- and third-growth forests and clear-cuts they usually work in. They struggled with a wraparound and rewrap method with the tape, finally measuring the fir at 169.4 centimeters around—almost five feet, four inches. It towered 66.4 meters—a Doug fir more than twenty stories high. Simard settled at its base, looking over her notes from her soil sample. She was quiet. Then said: "It will probably get logged. Anything in this district can be logged. And it will be."

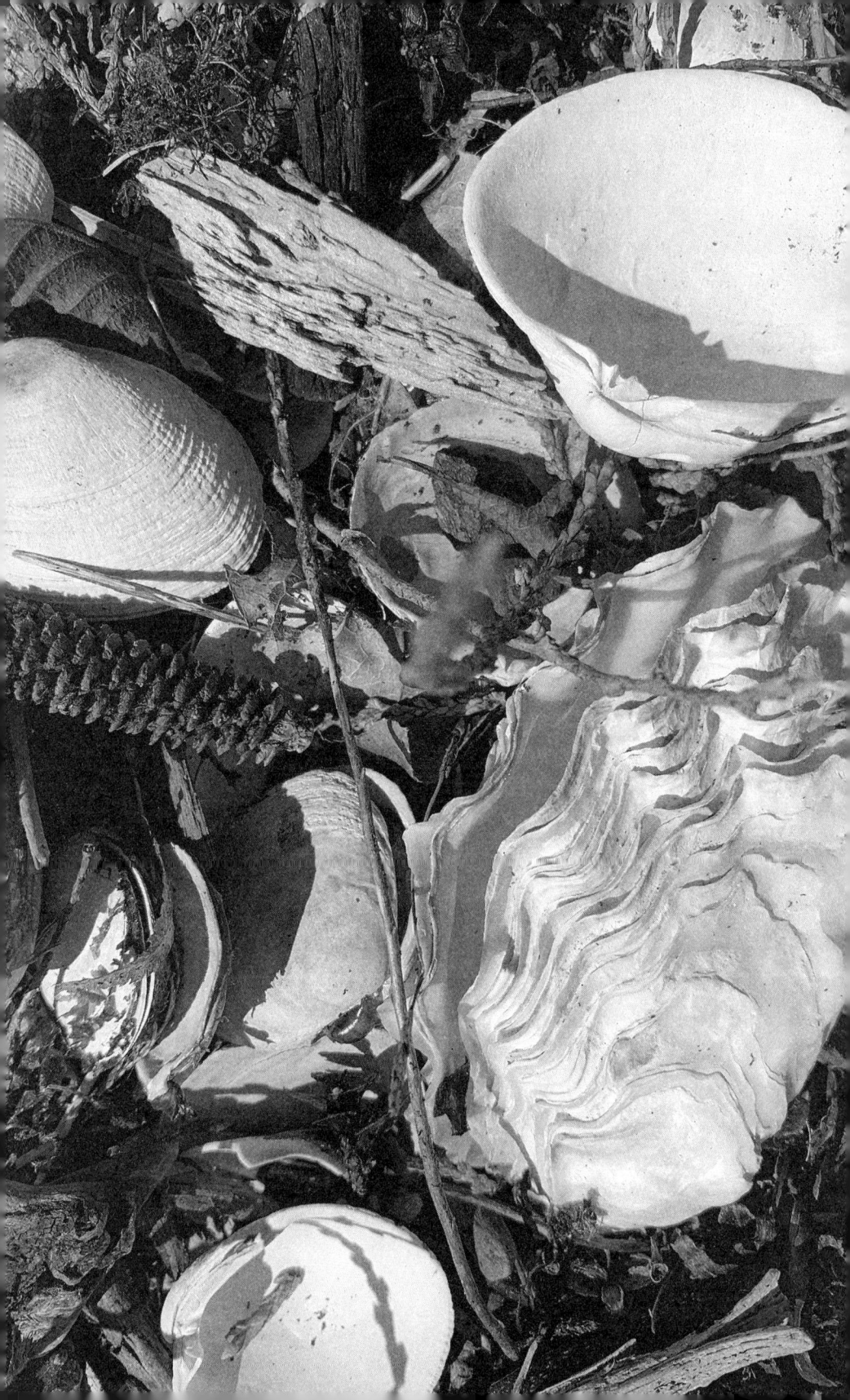

5

BIG TREES MATTER

AFTER MY WEEK with the Mother Tree crew near Woss, at the center of Vancouver Island, I headed north, to see where this all started: Nootka Island on the west coast of the island, the so-called birthplace of British Columbia. The newcomers were not the first to make use of these lands, as their stories would have it. Nor was this land pristine, an untouched wilderness, a terra nullius unowned or ungoverned by anyone, when the invaders arrived. No. The ancient trees and land itself here tell a different story, of societies thriving here for thousands of years.

The Nuchatlaht are one of more than a dozen Nuu-chah-nulth peoples. Their Nuu-chah-nulth traditional territories and family ties extend to Nootka Island and the Makah Reservation in Washington State, the northern boundary of which is the Strait of Juan de Fuca. At the time of settler colonization in 1846, they lived by complex and deeply held cultural protocols as to how the landscape would be occupied and its wealth used. Standing for centuries, the old-growth trees here are witnesses, unimpeachable in their testimony against the false "first-ism" and pristine narrative of what and who was here before. Long before the colonizers showed up, the First Peoples lived here, taking care of, managing, and using these lands. They planted, weeded, transplanted, coppiced, pruned, fertilized, and harvested forest gardens, abundant with crab apples, berry patches, and wild rice root crops. They used controlled burns to keep lands open and built stone fish traps at the shorelines to selectively harvest seafood. And always, everywhere that cedar grew, they utilized this Tree of Life for so much of what they needed, from birth to death, according to protocols and gathering rights that ensured the cedar trees would persist for centuries.

Scars on these trees and remnants of these forest gardens are readily seen today, for those who know how to look. New archaeological examinations of these landscapes are exploding old misconceptions of "hunter-gatherers" only lightly touching this land at coastal margins as they happened upon abundance readily to hand. "The whole concept of hunter-gatherer is useless, a stratified suffocation to put people in a social and evolutionary box," said archaeologist Chelsey Geralda Armstrong, assistant professor in the Indigenous Studies Department at Simon Fraser University in Burnaby, British Columbia. "I've never heard an elder say 'I'm a hunter-gatherer.' Take the term away." Geralda Armstrong and her coauthors published a paper in May 2022 that exploded the hunter-gatherer myth—and lets the land speak eloquently as to what and who was here before. Working with the Native people who have long lived in the study area, botanists and archaeologists identified ancient food forests, some with trees hundreds of years old. Here were edible fruit and nut trees, shrubs, berries, and roots. Wetland crops, berry patches, and orchards all continue to grow near archaeological village sites today on Nootka Island. No one should be surprised.

Naturalist Archibald Menzies, a surgeon on George Vancouver's voyage in September of 1792 while at Tahsis Inlet, on the west coast of Vancouver Island, in Mowachaht and Muchalaht territory, noted intensive cultivation of root crops tended "with as much care and assiduity as if it had been a potato field." He was referring, Geralda Armstrong explained, to intertidal root gardens. These are gardens at the meeting of land and sea, managed to encourage root crops including Pacific silverweed, rice root lily, and springbank clover. Just as intensively used were patches of blueberry, salmonberry, camas, salal, red elderberry, and orchards of crab apple. The First Peoples' use and management of an area, Geralda Armstrong noted, "did not start and stop at the longhouse." These remnant forest gardens and orchards still persist more than two hundred years after Captain James Cook's visit, despite government displacement of the First Peoples and extensive industrial logging. The false first-ism and terra nullius claims by Canada in taking this "abandoned" land for the Crown crumbles in this evidence, Geralda Armstrong and her coauthors wrote in their paper on forest gardens. The work of documenting this evidence was solicited by the Nuchatlaht Nation, a tiny band of fewer than two hundred people, for use in their case to the Supreme Court of British Columbia to reassert their right and title to their land. It is the trees themselves and these forest gardens that were among their witnesses, testifying to long stewardship of this place. Contrary to British Columbia's provincial lawyers' argument that the land was empty or abandoned, thus it was the Crown's for the taking, this evidence showed the long presence of the First Peoples prior to colonization.

The trees and forest gardens also are direct evidence against the erroneous misconception of the so-called hunter-gatherer. "It is so deeply engrained, it's treated like an evolutionary, biological category. It is not," Geralda Armstrong said when I spoke with her about the team's research. The term is not benign, she stressed, as it ignores the tenure and care over millennia of the First Peoples in their lands— never ceded. "What ends up happening is the erasure of people from the landscape. Hunter-gatherer is ephemeral and opposed to farmer." But, Armstrong and her collaborators found, these are deeply modified landscapes. "And to an incredible extent. Pruning, coppicing, weeding, digging, transplanting. This is permaculture." The botanists and archaeologists documented more than eight thousand trees deep in Nootka Island forests scarred by bark harvest by Native people— and that was with only 5 percent of the Nuchatlaht's land base surveyed. The dense concentration of archaeological sites, cultivated lands, and bark harvest sites shows not only that the Nuchatlaht occupied, managed, and used their territory but the practice of their worldview of *hisuk-ish tsa'walk* (everything is connected). Far from just stumbling upon abundant resources, this society cultivated and took care of what took care of them. The trees, the plants, the nonhuman beings they regarded and treated as kin, so important were they in their culture—and they remain so today.

Cedar was the Tree of Life, providing bark for cordage and fishing line, woven clothing, hats and capes, baskets, wood for house planks and roofs, boxes and chests, masks, and, when entire trees were taken, canoes. Jacob Earnshaw is an independent archaeologist based in Victoria, British Columbia. He went out in the field with some of us on the Tree of Life Project team, to show us how to recognize culturally modified trees (CMTs). Not far from the town of Campbell River, he pulled over to the side of the highway. Not fifty feet from the two-lane highway whizzing with cars was a living tree museum. Earnshaw ran his hand over the scar on a cedar tree, showing the telltale mark of where the bark had been stripped for use, at least 150 and perhaps as long as 300 years ago. The tree had healed over the strip, leaving growing lobes of wood nearly closing over the scar. The tree had kept right on growing after the bark was taken, in this careful method of harvest that took only a portion of the bark that was needed, leaving the rest. In this way, cedar trees could be used over and over again without killing them. Even whole planks could be split off one side of the tree, which would continue to persist for centuries.

These culturally modified trees stand out from the rest, if you know how to see them. "You are looking for a flat scar that goes from the base of the tree," Earnshaw said, pointing where the blade of a long-ago tool had been used to start the strip. He has seen individual trees repeatedly harvested in this way as many as nine times, continuing to live, in a carefully stewarded forest permaculture. Bark harvest was not random but permitted according to rights to harvest in a territory. "This places people on the land using resources, and it shows exactly where people were," Earnshaw said. Looking around us, we could see there were similarly harvested cedars everywhere. For the Nuchat-laht these culturally modified trees and forest gardens have provided

Bark scars document use by the First Peoples of the inner bark of this cedar, possibly for making clothing, baskets, or other household purposes. Taking only a strip of the bark does not hurt the tree, leaving it for use by future generations.

powerful testimony in a lawsuit they brought to the Supreme Court of British Columbia, to establish their long presence in, and stewardship of, their territory. According to Earnshaw, types of testimony from these culturally modified trees include stumps, remnants of canoe and plank manufacturing, and standing trees marked by plank removal and test holes—incisions made in the tree to test quality. The most common mark is the bark strip, with its telltale tapered or rectangular shape and tool marks from axes and chisels to make the initial cut. So far, some twenty-four hundred forest utilization sites have been recorded across Nuu-chah-nulth territories, from fifty-three thousand individual trees. Some show the marks of multiple harvest events.

The date of the harvest can be read by tree core analysis with precision to the calendar year. Using a borer drilled by hand into a culturally modified tree—or a cut stump—Earnshaw can extract a core, which when sanded and put under a microscope reveals rings made each year by a tree. That reveals the date of a bark scar in the tree's life too, because the scar shows plainly in the core. Sampling trees over a large area reveals the history of disruption of a flourishing culture. The pattern of trees analyzed so far in Earnshaw's work reveals a higher frequency of bark harvesting before the political upheaval, disease, and death that came with European settler colonization. In court, provincial lawyers can seek to discredit stories and oral histories. But the trees and forest themselves testify to the breadth of use and management of cedar, because it was so important for everything useful for life. In his work on Vancouver Island, Earnshaw has documented at least eleven hundred years of bark harvesting, with the oldest evidence of use revealed in clear-cuts.

Stumps don't lie: there, laid bare in the weathered wood, Earnshaw can see the shape of the healed scars from taking bark, long healed over by the tree, sometimes completely obscuring the scar from view from the outside of the tree. Clear-cuts ironically often offer views into the past obscured when living trees have covered over these scars. They are revealed, looking down on a stump, in cross-section.

Earnshaw noted one stump on the northeastern shore of Toquart Bay in Barkley Sound, an ancient western redcedar stump, nearly ten feet around, that was 1,165 years old at the time it was felled in 2011. The embedded bark scar in the stump showed the bark was first harvested when the tree was fifty-seven years old, dating the scar to AD 903. That is the oldest recorded culturally modified tree in the Americas and longest living culturally modified tree ever documented. This record is only beginning to be revealed, and needs to be protected, said Earnshaw, who thinks the majority of culturally modified trees documenting Indigenous forestry practices for at least the last half millennium lies hidden and unrecorded within standing old-growth trees—or is exposed and overlooked in industrial clear-cuts that have not yet been sampled for bark scars. Culturally modified trees "are likely to be far more widespread than currently known or imagined," he wrote in his article "Cultural Forests in Cross Section." "Industrial forestry continues an erasure of cultural forests and thus indigenous history on this landscape." The sheer number of culturally modified trees and depth of time recorded in use of the trees is opening a whole new understanding of the First Peoples' use of these landscapes, Earnshaw said. "You start to see the patterns, the number of times a tree or a place was harvested, the routes people are taking to get to these trees, where they are clustered." Also revealed is an active ethic of conservation. "The whole magic behind culturally modified trees is it's sustainable, permanent use," Earnshaw said. "Versus cutting it to take it all."

That didn't start until the arrival of the colonists, beginning with Captain James Cook, the first Englishman to set foot on what would become British Columbia, when he visited Friendly Cove on Nootka Island in March 1778. Yuquot—Where the Wind Blows from All Directions, in the Nuu-chah-nulth language—was called Friendly Cove by the explorers and traders. It was the sea otter trade that brought them to this place, seeking the fortune to be had in "soft gold"—sea otter pelts—the thickest pelage of any animal on Earth, with more than a million hairs per square inch. For the brief twenty years until the sea otters were hunted out, this once obscure and faraway place—still hard to reach today—was a center of world trade that linked Britain, India, and China as well as Russia, Spain, France, and the United States to this coast. Maxine Berg, professor of history at the University of Warwick in Great Britain, reported as much in her 2019 paper "Sea Otters and Iron." By 1820, several hundred European vessels had stopped on the coast, with the largest concentration of thirty-two vessels in 1792 alone, Berg wrote, and one hundred between 1778 and 1805.

The care for their lands by the First Peoples at Nootka Sound, the Mowachaht/Muchalaht people (who amalgamated in the 1950s), was noted by the explorers and traders, some of whom expressed irritation at being expected to pay for whatever they took, so sharp was the First Peoples' sense of ownership of their territory. Cook complained in his journals of the distinct sense of property the people firmly expressed, with regard to any aspect of their lands and waters: "Here I must observe that I have no where met with Indians who had such high notions of every thing the Country produced being their exclusive property as these, the very wood and water we took on board they at first wanted us to pay for." Far from a terra nullius, this was an owned, occupied, and managed place, replete with rules, protocols, and laws, and had been for thousands of years. These were established, wealthy cultures, thriving through whaling, river fishing, hunting, and foraging according to strict protocols.

Archaeologists document evidence of Mowachaht/Muchalaht people at Nootka Sound going back some five thousand years, but First Peoples' stories go back much farther. Long before the first Russian, Spanish, and British explorers arrived on the west coast of Vancouver Island in the 1700s. Even longer before there was a British Columbia, the Mowachaht/Muchalaht occupied and governed these lands and waters as their exclusive territories, according to traditional laws and values. Then, as now, every mountain, point, river, and fishing bank had a name, and the people of the territory knew its every corner, and the habits of all the animals and fish, the plants, the trees, and the birds. The people lived at Yuquot every summer *for more than five thousand years*, and traditionally every inch of the territory was represented by a chief. Under traditional law it was forbidden for anyone to use or access resources within each of these chief's territory, without the permission of the chief—and the guests were expected to pay tribute for the privilege. With the arrival of the colonizers came a change of worlds. They brought diseases that cut through the First Peoples like a scythe. Smallpox, malaria, influenza, typhoid fever.

This prosperous self-governed place of thousands of Mowachaht/Muchalaht people in the area around Nootka and Yuquot declined to a handful of survivors. The Mowachaht/Muchalaht estimate their population in the late 1700s to have been about 3,500 people. By 1865 the population was 365. By 1901 it was 247. Just 7 percent of the original community survived disease, violence, and dispossession. Yet a substantial Mowachaht community continued to live at Yuquot until 1966. Then the descendants of the Muchalaht people Cook and his crew encountered were forcibly removed by the Canadian government from their seaside shorelines inland to Gold River, on Vancouver Island. This displacement is still resented among the elders, the last of the First Peoples to go to school at Yuquot, to grow up, play, and live in what had always been their place. The Muchalaht hereditary leaders that run the tribe today persist in a continuous chiefly line from the era of Chief Maquinna, who in the 1780s so irritated Captain Cook, surprised to encounter a sophisticated trading culture and hard bargains. Today the Mowachaht/Muchalaht people are still led by Chief Maquinna, a direct descendant, and leader of the council of chiefs, staff, and citizens of the nation. The tribe is just as clear today as they were then on who owns these forests and their territory, never ceded—despite being declared "Crown Land" by British Columbia.

IN ADDITION to the demographic catastrophe and physical displacement inflicted by the colonizers, one of the most profound acts of colonial violence was the erasure of place-names. The Indigenous names
for every sort of place in the territory had forever denoted places of
abundance, way finding, spiritual power, or natural phenomenon. This
place literacy depicted in names was expunged and replaced with
meaningless European names to honor distant relatives, wives, and
holders of power. Entire systems of intimate knowledge and living
based on relationships, respect, and reciprocity with nonhuman kin
were pushed aside. Just as invasive were the settler-colonial lifeways
of extractive capitalism—with disastrous effects for the native ecology and peoples of this place. In 1846, Britain asserted sovereignty
over what is now British Columbia, which federated with Canada in
1871. An Indian reserve commissioner visited the Mowachaht and
Muchalaht in June 1889 and, in just two days, decided on and laid
out eleven small reserves for the Mowachaht and six for the Muchalaht, including less than 1 percent of the land in their territory. These
lands were declared by the agent as "small, some but little used, and
worthless except as fisheries and as furnishing a quantity of timber."
Canada took the rest of these unceded lands, paying nothing for
them. No treaty was negotiated, no aboriginal title was ceded by the
Mowachaht or Muchalaht, no war of conquest was fought.

It was not long before the industrial-scale cutting would begin.
The invaders who came for the furs would be replaced by those who
stayed for the trees. They pushed aside the First Peoples to places
where they had never lived, and proceeded to do with their lands and
water things the First Peoples had never done. Traditional culture
of the Mowachaht and Muchalaht was based on core management
principles and elaborate customs and laws reinforced in potlatches
and rituals handed down from family to family through generations.
Central to these was the understanding that their communities' wealth
was kept not in paper bills and metal coins or a distant bank but in
the *living* forest—in its trees, its waters, its animals, its medicines, its

spirit powers. The bark scars on trees that were stewarded to be kept alive and productive through centuries are testimony to an ethic based on long-term, place-based wealth. Now, suddenly came an alien land use, based on reaping individual profit and moving on to the next place. Cut and run.

IN THE SUMMER OF 2022, I went to Yuquot (now a National Historic Site of Canada) on the remote northwestern coast of Nootka Island, to see how it had fared over more than 250 years of colonialism. Roger Dunlop, then the lands and natural resources manager for the Mowachaht/Muchalaht, piloted the tribe's fisheries boat to Yuquot. It was a perfect August day, the sea wind tangy, the water green glass. Rocky islets were hatted with forests and porpoises knifed the water. Trees along the shoreline were sculpted to the blow of the wind. The mountaintops were quilled with old-growth trees, their silvered and broken tops spearing the forest canopy and giving the land a porcupine back. The water was so many hues: turquoise, aquamarine, jade. The boat threw rainbowed spray, and a morning haze clung to the water. Bleached driftwood was racked on the shore in great battlements. Dunlop pointed to a distant beach, where a yearling black bear worked the waterline, turning over rocks, perhaps in search of breakfast. He shouted over the sound of the motor, orienting me as we sped along.

"People are so lucky they don't understand what they are seeing when they come to beautiful Nootka Sound," Dunlop said. He pointed to the shorter, lighter-green forests that stood below the higher, darker-green patches of old-growth, with the silvered tops broken by age and weather. The cutover forest dominating the landscape but for the mountaintops. "This is all basically industrial woodlot," he said with a sweep of his arm at the passing shore at the northwestern end of Vancouver Island, now cut two and three times, leaving an ever-diminished forest. At the current rate of harvest, all of the old-growth left in Mowachaht/Muchalaht territory would be cut in the next fifteen years. "Gone," Dunlop said, drawing a finger across his throat for emphasis. Wild salmon populations also are in deep trouble, he said, headed for extinction in twenty years without extensive intervention. We were nearing Nootka Island. Dunlop shut off the motor, and Mowachaht/Muchalaht hereditary chief Jerry Jack Jr., our guide for the trip, hopped up on the dock at Yuquot, as Dunlop threw him a bowline to wrap around a cleat. Jack motioned for us to stay aboard as he took a moment, singing a song of blessing on this land, never ceded, before we came ashore. We quieted, listened, understood who this visit was for. Not only us, but Jack's ancestors, whom he had come home to see. At Jack's signal we walked up the boardwalk and onto the beach, a gentle curve of pebbles meeting clear saltwater.

There was a time, not long ago, that this shoreline was crowded with longhouses. Today only one family lived here, in a simple cabin. The only dwelling at Yuquot, in addition to a Coast Guard base across the bay, on the point. "We call this ground zero of first contact. It all started here," Jack said. "We held all the feasts, the potlatches here, in our big houses," he said, referring to the potlatch houses where ceremonies and gatherings would be held for days. "We are bitter about contact because of the way it changed our people. There were thousands of us here. Now it is one family. We have to prove right and title to our lands—it is really frustrating, the government saying the land doesn't belong to us." Jack's great-grandfather was put in jail for potlatching, and his father was taken away from Yuquot when he was five. Jack grew up in Gold River and eventually moved away to work in law enforcement. "We didn't get crazy until you guys showed up. We went from living off the land to living on a cell phone. So much changed so fast. We have been through a lot."

Jack, Dunlop, and I crunched along the pebble beach and settled on a driftwood log. Ravens talked in the deep forest and the tide whispered to itself. Fireweed waved its lavender wands from a meadow on the mainland. It was quiet. Summerfest, an annual community commemoration of survival when the families come back to Yuquot for a time of remembrance, had already passed. This is still the center of the world in Mowachaht/Muchalaht culture. Claimed as Crown Land and called the birthplace of British Columbia, Yuquot is persistently a deeply cultural Indigenous landscape and, unlike so many other places, even managed to keep its name. "Traditionally, this place is the hereditary chief's territory, with responsibility for everything that flies over it, swims through it, or lives under it," Jack said. Yet now the First Nation doesn't, in the government of Canada's eyes, own hardly anything here anymore.

Jack motioned for us to follow him up a path to the one residence here on the island, a simple cabin facing the sea. We were welcomed in and there, in an armchair by the window facing the sweep of sparkling blue, sat Ray Williams, then eighty years old. The last Mowachaht elder to live at Yuquot full-time, he had simply refused to leave. "Nobody was going to tell us to move, nobody can tell us where to live," Williams said, as I pulled a chair close to his good ear. "I love living along the ocean. I love living here because our food supply is here. When Indian Affairs moved our people [to Gold River], there is nothing there. You have to get it from the store. But it is not the same." All his life Williams had lived with and honored the surrounding forest. "It helps protect our medicine, and it helps to keep the rivers intact, to log out the forest it wrecks the salmon beds." Cedar is vital to their way of life, Williams said, for making everything from canoes and boards and longhouses to kindling, fire starter, and the bark for dance skirts, for wrist and head bands, everything you can think of.

One of the last to speak his language, Williams's teachings came from his elders. "Everything is connected," he said of his people's territory. I asked him about how the land had changed during his time in his territory. "When the trees were taken away, it started taking the pebbles away where the salmon were spawning. The logging destroys the salmon beds and scours the river. I've been sad for quite some time, even though I used to be a logger in my young days, I didn't

Elder Ray Williams, then eighty years old, visits with me at his cabin at Yuquot. Williams was the last Mowachaht/Muchalaht First Nations elder to live year-round at the ancient village site of Yuquot on Nootka Island, British Columbia. Photo by Erika Schultz.

think about what we were doing because we were young, and it was a job." He still remembers loading a logging truck with just one log, it was so big. "I didn't think anything, oh, it's a monster tree, big tree, the first thing our loggers thought, taking away that big fir log. It still hasn't changed, they are still taking the big trees. I was part of that, holy shit, what did I do anyway? I do regret it. I actually helped the forest company to get richer and making us into slaves for the forest company, making money for gas, for food. I want to tell what I can while I am still aboveground."

Our visit on Yuquot had come to end, and Dunlop herded us back in the boat. As we spun off through blue-green water to the north end of the island, he pointed at the blinding sparkle in the distance to the bobbing heads of sea otters, playing in the waves. He wanted to show me Owossitsa Creek, the place that inspired Salmon Parks, the Indigenous-led conservation effort launched by the Mowachaht/ Muchalaht and Nuchatlaht, to save these forests from more industrial-scale logging. Adopted as policy by all the Nuu-chah-nulth Nations, the intention is to set these lands aside for all but light subsistence use, let these cutover forests recover, and protect the rest, allowing these lands to heal and regain their natural rhythms and processes. Within the Salmon Parks, industrial harvest would be greatly reduced, by establishing a network of connectivity corridors for wildlife and forest recovery, so these forests may once again nurture the salmon, the people, and nonhuman beings.

These are truly ancient landscapes. Archaeological evidence has documented Chinook salmon and grizzly bears around nearby Little Woss Lake Basin on Vancouver Island dating back almost fourteen thousand years before present. Other studies have shown some of the north-facing slopes of Vancouver Island have not burned in more than six thousand years. Yet industrial cuts of these forests are made every sixty years now—increasing the natural rate of disturbance of these landscapes by one hundredfold. Today, about 80 percent of the Mowachaht/Muchalaht territory has been logged. It was a stream at the north end of Nootka Island that spurred the Nuchatlaht court case, and the launch of the Salmon Parks movement. The decision to go to court came only after this last place was nearly cut. Dunlop pointed out the encroaching damage nearing the creek to the late hereditary chief of Nuchatlaht, Walter Michael—cutting that had endangered the sockeye salmon that by traditional law, only Michael was allowed to harvest. Dunlop's suggestion started it all: "Why don't we just say no?"

We motored on and came to a glimmering cove. Dunlop idled as close to the beach as he dared, and dropped the anchor. We waded to the shore in our hiking boots, the water a cold shock even in summer. Here, meeting the sea, was the crystal perfection of Owossitsa Creek, a sockeye stream tumbling to saltwater through the last intact old-growth forest in Nuchatlaht territory at the northern end of Nootka

Owossitsa Creek on Nootka Island is a vision of perfection. The Indigenous-led Salmon Parks movement started here, to save salmon by protecting the forests that nurture salmon streams.

Island. We walked up the creek, in its perfection, so wonderfully alive. Here was a root wad of a giant tree, tumbled across the stream, so big the broken roots splayed more than ten feet overhead. Everywhere, deep forest hung over the water, sun beams finding their way here and there through the shadowy banks to light the water, running cool and cold.

There was a mix of species: redcedar, hemlock, spruce, bigleaf maple, and a glorious thick understory. Salmonberry, huckleberry, salal, and swaths of sword ferns. Water bugs skated on the surface of the creek, and a tumble of hemlock needles and leaf litter scooted along the bottom in the gentle current—food for everything that lives here. The rocks were glazed and furred with healthy green algae and moss.

Logs fallen in the stream impounded pools too deep to wade, and the stream glistened as it poured over the dead and downed wood, making crystalline bubbles, oxygenating the water. As we walked upstream, exploring, baby fish scooted under the logs that armored the banks, finding favorite hiding places for resting and feeding. This was premier salmon-spawning ground, in a stream singing through a forest that protects it, making its way in a gentle gradient to the sea.

Looking at this splendor, it was obvious what was at stake. This old-growth forest, keeping company with all the lives it sustains, not only in the living trees today but the dead wood and snags and wood debris they will become in the streams. These big, downed logs and the roots of still-standing trees provide the stability for the creek and cohesion of the banks of the creek that are a keystone of good fish habitat. This root strength—and the large amount of woody debris the forest continually sheds into the creek—keeps its structure complex, with all the wiggle and waggle of a natural stream, continually changing its path to wind around, over, and through woody debris. This is what mountain streams do, continually moving gravel, sediment, and wood from high slopes to the sea. This gravel and sediment are what salmon use to build their nests, and the trees on the banks shade and cool the water and hold the soil. I have read so many books about this, so many scientific papers, so many restoration plans underway at such huge cost to try and restore even a bit of the miracle of places like this, and their intertwined, self-perpetuating perfection.

Big old trees are crucial to the natural hydrology of the forest and streams that thread through them, scientists such as Julia Jones have shown. Young plantations suck up water, altering the hydrology of streams for decades, decreasing summertime flows, when water is scarce, while in winter, the cutover slopes allow a deluge of runoff into the creeks. Gone is the moderating influence of a mature or old-growth forest, that with its intact canopy and deep soils ensures a slow, steady steep of moisture into the ground and groundwater. This is the water that the forest moves up to the surface in summer with the hydraulic lift of its roots—a perfect flow regimen in concert with the wet and dry seasons of the year, moderated by the forest that is supposed to be here. When it's been cut, all of that natural stormwater management in winter and irrigation in summer is altered and even entirely lost. This natural movement of the water over the land is one of the most key benefits of big trees and can only be obtained with time. A lot of time. It's one of the first things lost with clear-cutting. As Franklin and his collaborators at the Andrews documented, natural forests are valuable not just for wood, but for the natural processes and ecosystem services they provide that take time—several centuries—to establish.

Another factor impacting the natural movement of the water over Yuquot is the mostly vanished beaver from this landscape, as they were intensively hunted on Vancouver Island in the same rapacious fur trade that did in the sea otters in the late eighteenth century. Their dam building in natural forests created extensive networks of ponds and wetlands that helped hydrate the land for thousands of years. All of these changes to the landscape on Nootka Island worsened conditions for fish, including depletion of the immense source of food that forests provide for aquatic life when forests are intact.

The Nuchatlaht were not entirely successful in their initial court case, although the judge in May 2023 left the door open for the Nation to come back and establish some of the title claims they sought. This they did, and won. In a historic victory for the Nuchatlaht the Supreme Court of British Columbia in April 2024 recognized the Nation's aboriginal title to more than four square miles of land on Nootka Island, including the lower reaches of Owossitsa Creek. With the ruling, the Nuchatlaht became the second-largest aboriginal title-holder in British Columbia. And the Nation was just getting started. The Nuchatlaht went back to court on appeal for acknowledgment of the entire original claim area, including whole watersheds, not just title to the narrow coastal strip recognized by the judge.

What a place the Nuchatlaht have already rescued for all time, with the success of their title case thus far. In the fall of 2023 the Salmon Parks vision got significant momentum, in a commitment of $15 million in funding for the Salmon Parks project from Canada's Ministry of the Environment and Climate Change to launch the project to protect old-growth forests and salmon in Nootka Sound. The money is intended to allow protection of nearly one hundred thousand acres of old-growth forest critical to salmon and all they sustain. Most of the money is for land acquisition costs, including buying out logging rights held by forestry companies and BC Timber Sales, the provincial forestry agency, on Crown Land. Already recognized under Nuu-chah-nulth law, the recognition of the Salmon Parks by the provincial government allows the areas to be protected by buying out the timber rights. Money will also pay for legal and administrative costs and even guardians from the community to monitor and report on the protected areas.

Dunlop and I followed Owossitsa Creek to the beach, climbing under and over the profusion of downed wood that creates the stream's meandering complexity and fish-friendly pools and side channels. I saw wolf tracks in the black, muddy bank, not far from where skunk cabbage was unfurling its wide waxy leaves. As I neared the delta where the stream wove itself into channels cutting through the beach, I could hear the tide and saw the iridescent lavender blue of mussel shells and bright chalky white of clam shells pressed into battlements of logs where the stream met the sea. Clamshells were piled on the beach, and heaped in shining drifts beneath the incoming tide, their gleaming profusion telling of the bounty of these waters.

Only nature can heal these forests, over time, with the return of the natural ecological function in these watersheds. That in turn will return access for the Nuchatlaht to healthy foods and medicines they have always depended on from their forestlands and waters. While on Vancouver Island, I met with Archie Little, a Nuchatlaht elder and councillor for the Nuchatlaht people who fought for the title case and supports the Salmon Parks. This conservation initiative could

Nuchatlaht First Nation elder and councillor Archie Little is helping to lead the fight by the Nuchatlaht to restore right and title to their ancestral territory, never ceded.

not only restore wild salmon by recovering key watersheds in Nootka Sound and the west coast of Vancouver Island, he said, but Nuu chah nulth knowledge systems, values, and guiding principles to this land. If successful, these efforts can undo a bit of the colonial intrusion, injustice, and resource exploitation and restore the principals of *isaak* (respect for all), *uu-a-thluck* (taking care of), *stalth* (together), and *hishuk-ish tsa'walk* (everything is connected). For these cosmological systems, if taken seriously, direct what should and shouldn't happen in these forests. "We have to respect Mother Earth, we have to heal it," he said.

Little is a survivor of Canada's government-sponsored religious residential schools, which removed Native children from their families, communities, and culture, ostensibly for indoctrination and assimilation into Euro-Canadian culture. According to the *Canadian Encyclopedia*, an estimated 150,000 children attended these schools, and thousands died there. Children were punished for speaking their language—and for no reason at all. I spoke with Little in a grove of old-growth trees in Cathedral Grove, not far from Port Alberni on Vancouver Island, in August of 2022. He and a generation of children are still recovering from what they and their families lived through in the boarding schools, he said. He recalled specific punishments as if they had happened last night. The rapes, the beatings, the blood. Thousands did not survive at all. "I am seventy-three years old, and I fear sleeping in the dark," he said. "They came at night."

Little believes the mistreatment of the forests and wetlands by the explorers, hunters, invaders, and settler-colonizers comes from the same violent root of dispossession and displacement of the Nuchatlaht's cultural knowledge. Here is a knowledge system that for thousands of years took care of these lands and waters, that while used and harvested, always were taken care of for the next generation. "Look at the First Nations, we survived for thousands and thousands of years, in healthy, prosperous governments and cultures," he said. "The non-Indians hitting our shores saw the wealth of the land and the ocean and the people. But it was community wealth. They saw personal gain, to take and take for profit. Not manage it wisely for the future." For this reason, taking the court case to British Columbia's Supreme Court was a matter of duty, Little said, to return the principals of the ancestors to the management of these lands. "We expand our duty beyond a buffer zone. That is looking after wealth. Once people understand how we are all connected to the forest, they will understand, we need to do better."

The Salmon Parks initiative is a five-hundred-year plan, launched by First Nations to put a new and different ending on the four-hundred-year history of colonial extraction in their lands that started on the other side of North America: in the *Atlantic* salmon forests of Maine.

6

MOSSY AND MOOSEY

THE AMERICAN NATURALIST Henry David Thoreau explored the Maine woods on three trips—in 1846, 1853, and 1857—traveling forests, as he put it, "mossy and moosey," the spruce dark "like a standing night." His descriptions of his travels are a gift nearly two hundred years later, for the witness Thoreau bore to a changing world. He wrote as extractive timber cutting took its first bite out of North America, in the country's most forested state, with 17.5 million acres of woodlands—90 percent of Maine's land base.

Pine was the most prized of trees. Coveted by the British for ships' masts, white pine twenty-four inches in diameter a foot from the ground was claimed for the Crown with the mark of the king's broad arrow—three slashes hacked in the bark—and a prohibition against cutting for anything but the Crown's use. Widely ignored by settlers, this policy nonetheless shows the wood famine of the Europeans, who had already stripped their own forests. So gigantic were these pines, they required special handling, with gullies filled, rocks removed, and streams bridged to even the surface where a tree would be felled, to prevent the tree from shattering on impact. A roadbed would be smoothed to pull the behemoth from the woods to the river for transport, with teams of as many as eight to twelve oxen and men pulling back on the log frantically on the descents. As late as 1850, one log containing 6,500 board feet was transported into Belfast, scholar David C. Smith reported in *History of Lumbering in Maine*, hauled by fourteen oxen. It sold for $250, nearly $9,600 in today's value.

In his history of the Baskahegan Timber Company, former president and CEO Roger Milliken Jr. described how these magnificent trees made easy targets in the forest: "A mature white pine stands taller than any other tree in the Maine woods, overtopping the rest by as much as 50 feet. . . . Timber cruisers climbed pine trees . . . [and] from these crows' nests, they gazed across the tops of the neighboring forest to where another, a vein or a glade of pine stretched its feathering limbs against the sky." Eventually the Maine woods were stripped but for the smaller or unsalable trees. "I would have liked to come across a large community of pine, which had never been invaded by the lumbering army," Thoreau wrote, but to no avail. Loggers cut nearly all the big pine and by the 1860s were targeting spruce, the most heavily cut species, as the loggers moved on. It was a pattern that would be repeated, moving eastward parallel to the Maine coast and northward inland, from river to river, with operations centering first at Saco then the Penobscot. Between 1846 and 1857, when Thoreau made his trips to the Maine woods, the Penobscot River basin was the most heavily cut in the state. Here was Maine's largest river, the second largest in New England, with Bangor at the head of navigation for seagoing vessels.

Sawmills were everywhere, Thoreau recounted, with 250 mills on the Penobscot and its tributaries above Bangor, cutting millions of board feet a year. So heavy was the cutting it grieved him. "Think how stood the white pines . . . branches soughing with the four winds, and every individual needle trembling in the sunlight . . . sold perchance to the New England Friction Match Company. . . . The mission of men there seems to be like so many busy demons, to drive the forest all out of the country." Thoreau described the milling of massive logs at the mill on the Penobscot at Veazie, in the Lower Penobscot, with its sixteen sets of saws, where the trees are "drawn and quartered." His language of mutilation and vivisection was vividly intentional. Such operations were wasteful, reported Geoffrey Paul Carpenter in his article in Maine History, "Deforestation in Nineteenth-Century Maine: The Record of Henry David Thoreau." Wood was so abundant and the logs so large, millers did not even try to use the whole log, but instead only portions that could most readily be made into a salable product.

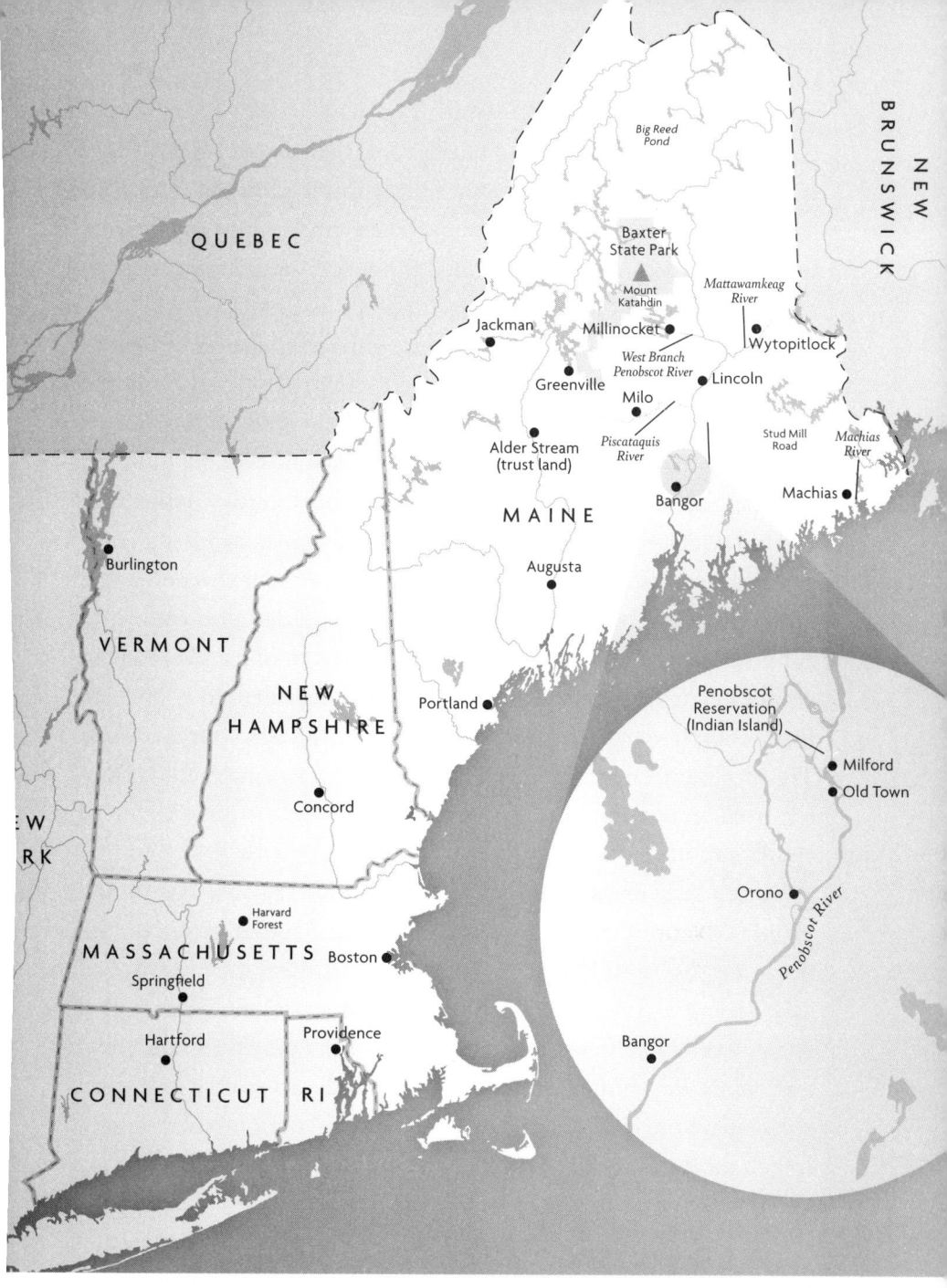

New England

Great quantities of scraps and tailings and slabs sawed from the logs were used to power mills or were simply dumped in the river. Raw logs went missing from log booms as booms broke, releasing floating logs cruising out to sea. Even milled timbers floated downriver from mill to market in rafts were warped, split, or just lost.

Wherever he traveled, Thoreau encountered the press of people, axes, and trade into the forests, and newly-cleared fields hacked into the woods, the logs still burning between hills of potatoes. The woods were rapaciously cleared for farming, for household heat, for building, for charcoal, and for smelting iron. Nineteenth-century settler families burned prodigious amounts of wood, as much as fifteen cords a year just for firewood. Forests were regarded as inexhaustible and in the way. Clearing practices were primitive. Settlers girdled trees to cut off their sap flow, then waited for the trees to die and fall apart, eventually leaving ground that was no longer shaded, so it could be planted. Or trees were cut, rolled, and burned, the ashes put in the soil. This slash-and-burn agriculture, in addition to logging methods that left tremendous amounts of slash and debris on the ground, fueled conflagrations.

By the time of his last visit to Maine, Thoreau saw loggers as a plague of vermin ransacking the woods, and a society that had lost, if it ever possessed it, any appreciation for a living tree for its own sake, seeing it instead as valued only as lumber or fuel. "Strange that so few ever come to the woods to see how the pine lives and grows and spires, lifting its evergreen arms to the light—to see its perfect success," Thoreau wrote in *The Maine Woods*. "But most are content to behold it in the shape of many broad boards brought to market, and deem that its true success. But the pine is no more lumber than man is, and to be made into boards and houses is no more its true and highest use than the truest use of man is to be cut down and made into manure.... A pine cut down, a dead pine, is no more a pine than a dead human carcass is a man." Thoreau saw a coarsening in a human nature to so destroy a living forest. "There is ... a higher law affecting our relation to pines as well as to men. . . . Every creature is better alive than dead, men and moose and pine-trees, and he who understands it aright will rather preserve its life than destroy it."

But it was not to be. The path to destruction that Thoreau lamented had not even really gotten started—and it did not stop at forest's edge at the riverbank. Far from it, the destruction of the forests was inextricably bound to the remaking of rivers as sluiceways for driving logs to the mills, with dams built to create lakes and ponds to hold the logs for the spring drive. This too saddened Thoreau, in the destruction of the forest caused by the rising lake levels, dammed for the storage and release of logs downstream. "They have thus dammed all the larger lakes, raising their broad surfaces many feet," he wrote in *The Maine Woods*, ". . . thus turning the forces of Nature against herself, that they might float their spoils out of the country. They rapidly run out of these immense forests all the finer and more accessible pine timber, and then leave the bears to watch the decaying dams." Thoreau decried the shorelines heaped with ruin, with its "maze of submerged trees, all dead and bare and bleaching, some standing half their original height, others prostrate and criss-cross, above and beneath the surface, and mingled with them were loose trees and limbs and stumps, beating about."

He described these same shorelines, ragged and unsightly and difficult to travel, in lake after lake. The impoundments were created to corral huge booms of logs, chained together in rafts many acres across. The logs were sent tumbling downriver in the spring flood, the jams that inevitably occurred blown up with dynamite or picked apart by men working the spring drive, using pikes to prod and poke and coax the logs as needed, riding the spring melt-out on bateaux. These log drives, for all their boisterous work songs and tales of daring, left a legacy as enduring as their lore, in deeply scarred rivers. Rivers were "improved" to make the drives easier by building dams, sluiceways, canals, slips, and removing rocks from channels by blasting. Day in and out, workmen split rocks and blew up ledges, cleared downed

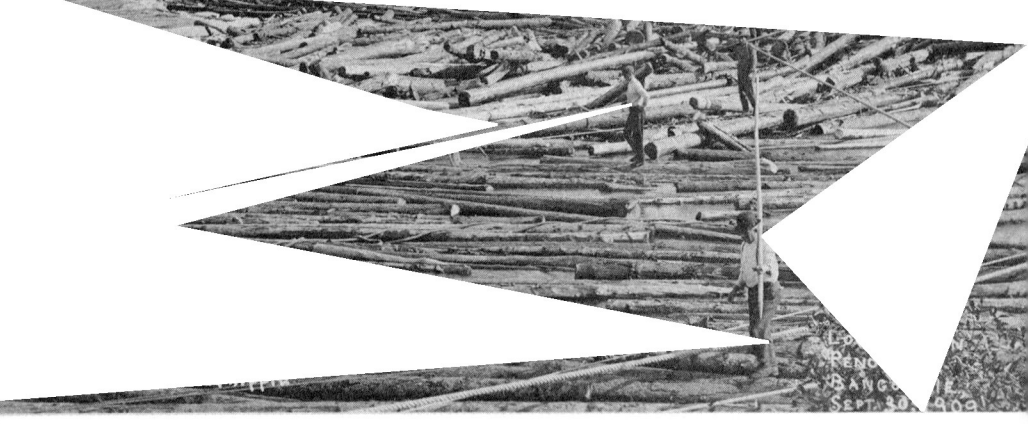

Log drives destroyed the habitat for salmon and other sea-run fish on the Penobscot and its tributaries. This log jam was photographed on the Penobscot at Bangor, Maine, in 1909. Photo by Leyland Whipple, Library of Congress.

fallen trees and driftwood and other woody debris, and removed gravel beds, with tolls levied to pay for it all, created by state legislative enactment. Hundreds of miles of rivers were simplified, smoothed, and shaved of their timber, their waters diverted in sluiceways and canals, and impounded. Yet nature was still problematic: the spring freshet could be both good and bad news, with water high enough to float logs to the mills—or so violent it broke booms and even swept away mills and sent logs and lumber floating out to sea.

THE REAL DAMAGE was yet to come, with modernization of the logging industry. The first railroad lines into the Maine woods were chartered in 1832, between Bangor and Old Town, made of wooden rails with strap iron on top. By road, by rail, and by river, trees were stripped from the woods and brought to the mills and the lumber sent around the world. By 1820, 746 sawmills were listed in Maine, with forty-two in Bangor alone, more than anywhere else in the state. Penobscot County was the heart of the action, with sawmilling establishments of all kinds in river towns just north of Bangor: Milford, Old Town, Orono, and Veazie. Ships' spars, masts, and knees, even hemlock paving blocks for the streets of Boston and cordwood for the wood-starved cities: all were cut, made, and supplied from the trees of the Maine woods. The railroads alone burned thousands of cords of firewood in a year's time. In addition to boards and timbers, an endless list of other products was made from wood cut and milled in Maine, from shingles, clapboard, staves and heads for barrels, hogs-heads, casks, and pieces for wooden boxes, to window sashes and doors and blinds in addition to everyday items. These included the handles for rakes, shovels, and hoes and bobbins, spools, spokes, and hubs. Excelsior for packing, shoe pegs, coffins, tubs, kegs, and oars.

Much of what was needed in life from birth to death came from the Maine woods and was milled on Maine rivers. Hemlock was also logged for its bark alone, feeding more than 248 tanneries work-ing by 1820 in the state. The felled trees often were left to rot. The Maine woods also were supplying the world's wood, whether boxes for Cuban sugar or lumber for the West Indies. By the beginning of the nineteenth century, Maine was furnishing about three-quarters of the pine exported from the United States, with hundreds of loads of lumber sledded out of the woods each day. In 1825 twenty-five vessels were transporting lumber from Belfast, Maine, to the West Indies. The shipping ports of Bangor, Bath, Calais, Ellsworth, Ma-chias, and Portland all were busy with the export of Maine's forests in ships loaded with clear and planed boards, paving blocks, sugar box pieces, shingles, clapboards, pickets, laths, ships' knees, cords of tanning bark, and more. The amount of woods stripped, milled, and shipped was colossal; one concern alone, S. W. Pope and Company, sent more than 3 million board feet out of Machias in 1858. Imagine my surprise to be walking across a bridge in Machias and see a bronze plaque set in the concrete, memorializing the Pope family for their prodigious industry. The same family that, after cutting these forests, set up their mill in Port Gamble, Washington, running from 1853 until

1995, one of the longest continuously operating sawmills in North America, cutting the forests of Puget Sound. The history of lumbering in Maine was repeated in the Pacific Northwest, sometimes even by the very same families.

By 1842, Bangor claimed bragging rights of shipping more lumber than any other place in the world, mostly to Cuba, Europe, South America, and the West Indies. This overseas trade was in addition to a busy coastal trade to other states. Bangor would become the largest lumbering center in its day anywhere, sawing 250 million board feet of lumber in 1872, the city's all-time peak production year. But by then, Maine's primacy as a lumber capital was already fading. All the big pine and spruce were already gone, and the virgin forests and new mills in upstate New York and Michigan beckoned, according to historians David Dobbs and Richard Ober in their book *The Northern Forest*. The Maine woods would never be the same. By the 1880s the eastern wolf, mountain lion, beaver, American marten, Canada lynx, caribou, otter, wild turkey, puffin, great auk, Labrador duck, and passenger pigeon were largely extirpated from Maine. Black bear, white-tailed deer, moose, Atlantic salmon, alewife, bald eagle, osprey, common raven, pileated woodpecker, and great blue heron were in steep decline as their habitat was stripped for lumber. It wasn't just the logging but the state bounties that culled the animals from the woods as wolf, mountain lion, bear, bobcat, and lynx were targeted. The bounty on lynx, today a federally listed threatened species, was not lifted until 1967, Andrew Barton and his coauthors report in *The Changing Nature of the Maine Woods*. Gone along with these animals were their key roles in the landscape, whether as top predators, grazers, or seed dispersers. Beaver created the ponds and wetlands that hydrated the landscape.

And yet with all these woods and rivers had already seen, the most profound and lasting damage to the Maine landscape was yet to come. The pulp and paper industry transformed the timber industry, landscape, and rivers of Maine, with even greater environmental damage. With the discovery in 1888 of sulfite to make pulp from softwood trees, companies such as International Paper of New York started buying huge tracts of land from the failing lumber companies and farmers along the rivers of Maine. Suddenly tree size didn't matter anymore; any tree was worth money as long as it had enough fiber to grind into chips, dissolve into pulp, and press into paper. According to the economist Lloyd C. Irland, in the early 1890s the Rumford Paper Company ran the world's largest newsprint mill—only to be eclipsed by the Great Northern Paper Company, beginning with construction of its mill at Millinocket in 1899, followed by the acquisition of an existing mill in Madison, Maine, and the construction of a third mill in East Millinocket in 1906. The Great Northern Paper Company developed the first paper machine in the world, Irland reports, to run at one thousand feet a minute.

It is hard to fathom the extent of the paper industry at its peak in Maine. Great Northern Paper became the largest consumer of logs driven down the west branch of the Penobscot River, to feed what at the time was the world's largest industrial plant and consumer of wood—powered by the largest privately-owned hydroelectric dam anywhere. A tourist postcard from 1908 shows the log pile at the Great Northern Paper in Millinocket with a mountain of twenty-one million logs. Historical photos show the river choked with wood, bank to bank, for miles. From 1969 to 1972, the Golden Road, cut through the woods for ninety-seven miles parallel to the west branch of the Penobscot, was constructed by Great Northern Paper to get at its 2.1 million acres of forest in the North Maine Woods to feed its mills in Millinocket and East Millinocket. It was the passage of the Clean Water Act of 1972—and the completion of the road that same year, allowing truck hauling—that brought an end to the log drives, outlawed by the state legislature in 1975.

Cutting intensity rose sharply particularly on spruce, which produces the long, strong fibers sought for papermaking. By the 1890s some of the hardest cutting the Maine woods had ever seen was well underway, with the smaller trees left from the earlier lumbering taken now for pulp. While the forested area of Maine increased following the abandonment of farms, harvesting of first- and second-growth was intense, at about one hundred million cords over forty years between 1880 and 1920, or five times the harvest taken in the previous 250 years, Barton reported in *The Changing Nature of the Maine Woods.* Along with the increase in cutting came the consolidation of the pulp and paper industry controlling mills, timberlands, and water in vast private holdings. The harvesting of large pine and spruce with men, oxen, horses, and rivers that so distressed Thoreau was romantic compared with the mechanized systems of high-efficiency cutting that were to come. Mechanization began with the motorized vehicles of the 1920s and 1930s and railroads punching into previously inaccessible areas. In the 1950s the chainsaw began to replace the crosscut saw, and by the 1960s skidders had replaced horses in hauling out the cut. By the 1950s nearly all sizes and species of trees were cut, and huge companies controlled nearly all the cut to their specifications of four-foot-long logs for pulp.

The environmental destruction endures today. "You take the forest away, and what happens?" said Dan McCaw, fisheries program manager for the Penobscot Indian Tribe, as we stood by the Penobscot in the winter of 2022, talking about all this river has seen. He continued:

Now the earth isn't held by anything, and you get massive sedimentation. There is nothing to buffer it anymore. The water sheets over that landscape and hits the river with a torrent. Ten thousand years of dynamic, stable equilibrium is out the window, just gone. And that is the very basis of everything. Everything comes from the form and function of that river, and the movement of sediment and nutrients in that river. And you blow it all up, and you do it on the main stem, and then on every tributary, and on every headwater. Cut the trees. Straighten the river. Bulldoze it. Even in the remote places, you see the results. And every animal that spent its entire career and its ancestors' career in that watershed, all that genetic wiring, that was thrown to the wayside, because now that river system, you can't count on it anymore. Then you move forward to the really horrid industrial revolution, now the dams that used to just run sawmills and grist mills and shingle mills, they are on everywhere on every little tributary. Now as the technology grows and machines grow, we can get into the bigger rivers, and into the main stem. You get the Great Golden City in the Woods. Millinocket. It's "look, guys, cheap labor, an endless supply of forest land, and a government willing to give us anything we want."

By the early 1970s, as Irland reported in *The Northeast's Changing Forest*, haul roads were pushed into the northern backwoods: "The silent fir flats of 'black growth' and the peaceful lakeshores were invaded by massive diesel log hauling trucks, carrying huge loads of tree-length wood." Total harvest increased one-and-a-half-fold, with paper production and pulpwood harvest surging ahead of lumbering. By 1988, 60 percent of all harvest by industrial landowners in northern Maine was by clear-cut. Public outcry at forest practices that had not been seen before in Maine led to a series of reforms beginning in 1989 that did reduce the size of clear-cuts but had perverse consequences. Harvesters avoiding the new requirements of the law started fragmenting the forest into strips and dice. The clear-cuts were smaller, the volume of wood about the same or a bit more—but the amount of disturbance, fragmentation, and edge cut into the forest doubled. The introduction of so-called whole-tree harvesting left not even slash to nourish the land and help it recover. Northern Maine had become home to an extreme version of industrial logging, with the entire north half of the state almost entirely held by a few industrial owners using the land and waters for pulp and paper manufacture. Nearly half the state's forestlands, 8.1 million acres, was owned by eight Fortune 500 paper corporations and by sawmills or other manufacturers, Dobbs and Ober wrote in *The Northern Forest*.

Maine is really two places, with the money and population in the south while the northern and mountainous forested areas of the state, now largely owned by multinationals and foresters, on perhaps the sixth cutting of what remains a largely forested landscape. It is an almost unworldly place. The so-called Unorganized Territories of the North Maine Woods with no municipal governments, few people, and a maze of logging roads became a de facto plantation run by the paper mill owners for tree cutting to feed their pulp mills. The wildlands became the land of square-mile clear-cuts, visible from space. "All across central and northern Maine, entire townships were being reduced to thickets as the paper companies boomed," Dobbs and Ober wrote. Jungles of raspberry, poplar, and scrawny gray birch covered the land managed for quick cash rather than its long-term value.

But yet another change in the fate of these forests was yet to come in a change in tax laws and ownership that would make ecologists and workers alike yearn for the paper mill days, which seemed cozy by comparison, in a paternalistic kind of way, as author Jamie Sayen described in *Children of the Northern Forest*. This was the same change world-renowned ecologist Jerry Franklin decried in the forest ownership of the Pacific Northwest: new tax structures favored forestland ownership by timber investment management organizations (TIMOs) and real estate investment trusts (REITs) rather than timber corporations. Suddenly, instead of mill owners maintaining a steady cut for mills and workers they employed in a vertically-integrated chain of production, came a sell-off of timberlands to owners with nothing to do with forestlands, mills, or workers. The commodification of nature that began with the arrival of the first settler-colonizers was now complete, with a series of changeovers of landownership and subsequent liquidation cuts by hedge funds, endowments, foundations, pension funds, and other distant financial owners with no connection to the forest other than by the quickest possible return on their investment on purchase of the land—preferably within seven to ten years.

Beginning in the 1980s, but mostly occurring between 2000 and

2008, the forest products industry divested itself of the vast majority of its timberland, with most of it going to REITs and TIMOs, corporate entities without any long-standing ties to the land they owned or the people who lived and had relied on that land for employment, recreation, and subsistence needs, wrote Andrew Gunnoe of Maryville College and his coauthors in "Millions of Acres, Billions of Trees," published in *Rural Sociology* in 2018. These new owners had no relationship with the local communities or timberlands they controlled, or even a corporate name or reputation to uphold on the lands they managed—so invisible and distant was their ownership. These changes had affected more than fifty million acres of Maine timberland, as corporations in the US forest products industry either sold off most of their timberland or restructured themselves to take advantage of changes in the tax code. It was the largest transfer of landownership in the United States this century, valued at more than $40 billion.

This was the culmination of a progressive distance and severing of connection between the people who live and work in the forest and mills they fed, from individual absentee ownership to corporate ownership, and faceless institutional financial ownership of a commodity that just happened to be a forest. In this thinking, a forest was conceived of as pulpwood factory, a fiber farm, biomass, or simply a commodity to be bought or sold with return on capital for shareholders emphasized over any other value. The TIMOs and REITs stripped away any last relationship with forests, which could be liquidated to increase short-term shareholder return. This was the land of the liquidation cut, an ultimate financialization of nature. From 2000 through 2009 most stands were cut at the rate of some half-a-million acres per year for pulpwood, sawlogs, biomass, or firewood—the highest harvest rate reported in state history, Barton reported in *The Changing Nature of the Maine Woods*.

These problems were not limited to Maine. In an article in *Forest Ecology and Management*, John Gunn of the University of New Hampshire and his coauthors found that by 2018, almost half of the forestland in Maine, New Hampshire, and Vermont was in degraded condition due to past management: young, small, and understocked in desirable species of good-quality sawlog trees, in part because of relentless commercial clear-cutting and high-grading, exploitive cutting that takes all the best trees in a stand and leaves the rest. Also lacking was the complexity of big old trees, dead and downed wood, or the habitat that sustains the life that makes a forest a *forest*—from its lichens and fungi to its animals. Their animal homes had been fragmented or removed altogether.

THEY WEREN'T THE ONLY ONES to be hurt. By the 1990s the paper industry was shedding jobs, as capital chased the next cheaper opportunities for paper production in the Southeast and then overseas. Many paper mills in Maine closed and others were just hanging on, and the communities that depended on them were struggling. Mitch Lansky's 1992 book *Beyond the Beauty Strip* is still the definitive work on the transformation of this landscape—his home ground. Lansky's critique is not with the loggers—he worked in a sawmill himself as well as in construction and digging sewer lines to support his family with his wife, farmer and artist and Master Maine Recreational guide Sue Szwed. Lansky's life's primary work, however, has been to lobby, write, and argue against the pollution, environmental destruction, and social ills caused by unsustainable practices that have destroyed the health and productivity of the North Maine Woods and the people who live there.

In his book Lansky looked beyond the border of trees left along the highways and rivers of Maine to the cutover lands and what skinning the land has meant to the local people. "We had a PTA, a civics club, a women's club, a fish and game club. It all collapsed," he said over dinner at his home, when I visited in the winter of 2022. Lansky, a former town manager and select board member, said his close-knit community had unraveled along with the forest. Mechanization—

Author Mitch Lansky of Wytopitlock, Maine, has been a voice for the North Maine Woods since he began his forest activism in the 1970s against aerial spraying and massive clear-cuts. I took this photo of Mitch in cutover industrial forestland not far from his home in the winter of 2022.

and log exports from Canada—meant fewer jobs in a town where most of the men used to work in the woods. Lansky built his house and a small subsistence farm and with Sue raised a family in the community of Wytopitlock. It is a beautiful place. Wytopitlock lies along the Mattawamkeag River, and is an unincorporated village in Reed Plantation, comprising sixty square miles in Aroostook County, settled in the early 1830s. Its population has been dwindling, with 207 people in the 2000 census but only 129 in 2020 and 118 in 2022.

House after house along the road was boarded up, I noticed. So was the elementary school, with the chairs and desks heaped and moldering in a dumpster by the front doors. With just two people per square mile, the area did not have enough kids to keep the school going, and it closed in 2008. This part of the state has been losing people for decades. The ten million acres of the industrial forest in northern Maine remains largely free of year-round habitation. Today it is a damaged place; paper companies in the 1980s, 1990s, and 2000s built twenty-five thousand miles of logging roads, enough to circle the Earth at the equator, and clear-cut more than two thousand square miles of forest, an area the size of Delaware, notes author Jamie Sayen in his book *Children of the Northern Forest*.

Bombardment of Lansky's land—and organic garden—in 1976 with aerial spraying of Sevin 4 Oil (carbaryl) launched him into his career as an anti-spray activist, author, researcher, and critic of timber practices in the North Maine Woods. The spray, an insecticide toxic to a broad range of organisms, was unleashed by the state of Maine on swaths of timberland in an attempt to control a native insect, the spruce budworm. The spray program began in the 1950s with DDT and lasted through the 1980s. The same year the state spewed insecticide over Lansky's land, the state sprayed 3.5 million acres of forest. Eventually the state switched to Bt, a biological spray toxic to caterpillars.

On a cold, bright winter day, Lansky and I walked a mauled landscape in the timberlands he has been watching and writing about since he started looking at satellite photos of clear-cuts in the 1980s that went for miles, some even the size of townships, or thirty-six square miles. During his activist days Lansky put together slideshows of those clear-cuts west of Baxter State Park and along the Allagash Wilderness Waterway, as cutting rotations sped up to forty-year cuts followed with herbicide treatment to clear the way for the growth of the next monoculture planting. The impacts to the soils, the habitat fragmentation, the short rotations, the degradation of the forest with herbicide treatment—none of that was being talked about. The reform he sought ironically led to the strip cutting—nonselective partial harvest—which, as some practice it, goes beyond high-grading to take anything of market value and leaving the rest. Here, from a bird's-eye view, was not Thoreau's brooding spruce tall and dark as "a standing night," but the rumpled skin of strip cuts that from the air look like the whorls of a fingerprint. Trees were cut as small around as peanut

This moonscape is a clear-cut in the North Maine Woods. It was replanted twice but never recovered from clear-cutting and treatment with herbicide.

butter jars, to be chipped for pellets or other low-value products. The equipment to do this—mechanically cutting, limbing, and stacking the trees—had carved ruts a foot and more deep in the ground. Blackberry, goldenrod, poplar, and spindly gray birch grew in thickets on the land between and around the ruts. A satellite view from Google Earth suggested this was the condition of the forest for miles around.

I've seen plenty of clear-cuts, some covering entire mountainsides. But this strip cutting, with the smallness of it all—the ever-fewer jobs, due to mechanization, cutting ever-smaller trees made into ever-smaller chips in an industry reduced to shipping the forest away as chips, pulp, or pellets—seemed such a diminishment of a storied and vital industry and of a magnificent expanse of native forest. A still intact wildland of continental significance has been degraded, simplified, cut up, and is haunted by loss—loss of wildlife, loss of jobs, loss of population—loss of *forest*, converted to thickets of tiny trees clipped, chipped, and shipped from a corduroy land. Just as damaged were the towns whose lifeblood had been the paper mills these forests fed. As the paper industry contracted, these Maine mill towns had suffered too.

7

STINKIN' LINCOLN

PAST THE STATUE of the giant loon at the traffic roundabout, past the welcome sign urging "COME FOR THE LAKES, STAY FOR THE LIFESTYLE," looms Lincoln's dead paper mill. There are many like it along Maine's Penobscot River, once home to a thriving paper industry. Today this is a landscape of cutover forests and a rust belt of dead paper mills. The story of each of these mills, and the towns they supported, is different—and in so many ways, the same. "Unknown to a lot of people, we have what is almost parallel to the rust belt disaster in Detroit," said Rick Bronson, town manager of Lincoln, Maine. "The auto industry has shrunk in Detroit, and the paper industry has greatly shrunk here. It has shrunk for one hundred miles, from Millinocket to Bucksport." The mill at Lincoln, once the town's second-largest taxpayer and third-largest employer, shut down in 2015, after operating for more than a century.

The story of New England is the story of the continent's first round of industrial booms-and-busts. The timber and paper barons gnawed this place bare, learned how to make paper out of small logs when the big ones were gone, ran that to the ground too, then moved on. From the industry's peak it has nearly collapsed—from more than thirty-five mills in 1900 to six by the close of 2023. By then, there were some three thousand paper mill jobs in the entire state of Maine— fewer than once worked at just the Great Northern Paper Company's mill at Millinocket. It employed about forty-four hundred people at its peak from the 1960s into the 1970s, according to a timeline by the Millinocket Historical Society. The mighty mill closed permanently in 2008, its silent stacks demolished in 2014. Today trees grow through the railroad tracks that serviced the mill. By the side of the road in East Millinocket is a memorial to Great Northern Paper, grass growing up around two granite rollers used in the factory. Like a tombstone, a bronze plaque memorialized the company that had built this town. Great Northern's East Millinocket mill began operations in August 1907, the plaque noted, and was modernized and expanded multiple times, ultimately producing nearly three hundred thousand tons of paper a year on two machines, each the size of a football field. Running the machines at forty miles per hour, workers made telephone directory paper and newsprint for customers around the United States and the world. Farther upriver, the Millinocket mill and the dams on the west branch of the Penobscot were a grim reminder of how far an industry can fall and what a scar it can leave behind when it does.

A March 2022 profile of the paper industry by the State of Maine shows freefall any way it's measured: jobs, contributions to the state economy, or sales—with those trends expected to continue. The paper industry, like so many others, has largely moved offshore, chasing lower wages, lesser regulatory standards, and cheaper production costs. Consumer tastes have changed too, and products faded, with the rise of e-books and online advertising and decline of newspapers all denting demand. It says something that one of the industry's bright spots is an aging American population in need of incontinence products. Demographic trends in that market are looking good, the state reports. As the paper industry has declined and mills shuttered, towns like Lincoln that for more than a century loyally embraced their mills for the jobs and benefits they brought, have been left to deal with the mess. So has the Penobscot Nation.

IN THE SPRING OF 2023, Dan Kusnierz, water resources program manager for the Penobscot Nation, invited me to observe some work he was doing for the tribe on a contract for the Environmental Protection Agency (EPA), sampling fish for contaminants downstream of Lincoln

Abandoned pulp and paper plants such as this mill in Lincoln, Maine, litter rivers and towns along the Penobscot River from the Canadian border to tidewater. Cleanup is costly, and towns struggle to find a new economic future.

Paper & Tissue, the abandoned, long-shuttered mill in Lincoln. A thirty-year employee of the tribe, Kusnierz intimately knew the history of this mill. So much so that around the tribal offices he went by the nickname "Dioxin Dan," said with affection by his colleagues. Riding in his truck out to the mill from the tribal center, Kusnierz spoke of the times when the power went out at the plant, so industrial waste from its treatment plant went directly into the river. Or when the giant pile of bark with dioxin-tainted sludge in it seeped into the ground until the tribe took the operator to court to stop it. Or when pipes from the plant discharged directly to the Penobscot River. The plant is upstream of Indian Island, home to the tribe's main settlement.

Early pollution of the river was troublesome enough: sawdust by the ton, untreated sewage, and dumping of industrial wastes—

everything from the muck from woolen mills to potato starch and pea vines from frozen food plants. Mill waste from tannins, loose bark and sinkers, and sawdust were commonly discarded directly into rivers. It would get worse. Papermaking is a chemical-intensive business, and pollution continues today. The EPA reported in March 2023 that five companies—including four paper mills and one food processor—produced 92 percent of the toxic chemical releases in Maine—7.4 million pounds of toxic chemicals released into the air, into water, and onto the land. The ND Paper mill in Old Town, downstream from Lincoln, prior to its shutdown in April 2023, reported releases of four cancer-causing chemicals including lead. The levels reported to the EPA actually were an improvement, as the decline in the industry has lessened its environmental burden. The river has been a dumping ground for paper plant pollution for more than a century, including from the plant at Lincoln. "In the early nineties, it would be obvious where the discharge was," Kusnierz said. "The water would just be boiling and it would have a big sheet with clumps of foam, stained with brown, a foot thick." To tackle the foam, plant operators put a defoaming agent in the discharge, then discovered one of the largest sources of dioxin discharged to the river was the defoaming agent, Kusnierz recalled.

The dead paper mill looms at the edge of Lincoln, its caved-in buildings scattered over the site. As we drove onto the abandoned property, deer and turkeys, long used to having the place to themselves, scattered. While Kusnierz and a water resources field lab technician fished for suckers downstream of the mill in Mattanawcook Stream, I wandered the abandoned plant. It was early May and nature continued its unstoppable work of bringing new life even to this abandoned brownfield. An osprey with a stick in its mouth flew to a light tower, where it was working on a nest. Pigeons cooed from inside the ruins. Lincoln Paper & Tissue declared bankruptcy in 2014 and walked away from the plant after operating, through many ownership changes, since 1882. To pay its creditors, the owners had opened the millsite to scrappers, who tore through the roofs and walls of the buildings to rip out anything of value. Loose siding creaked and screeched as it flapped in the wind. There was debris everywhere scattered over the ground: a worker's glove, the shiny yellow plastic dome of a long-gone worker's helmet. A fire extinguisher. Pipes loose and akimbo clanged and banged. Everywhere inside the building were deep drifts of paper pulp, like dirty snow. A hard wind moaned through caved-in wire fencing around the plant. Something was making a sheen on the puddles.

The mill produced products of everyday life: the postcards that tumble out of magazines, trying to get you to subscribe. Paper towels, napkins, plates, and stationery. It specialized in deep-dyed tissue used for pleasantries: party streamers, paper napkins, and tablecloth sets. In producing these ordinary things, the mill discharged extraordinary

Dan Kusnierz, water resources program manager for the Penobscot Nation, has worked with the tribe to clean up industrial pollution in the Penobscot River for decades.

pollution into the Penobscot River: dioxin and furan, contaminating the fish, sediments, and other natural resources, according to a May 2001 filing in US Bankruptcy Court for the District of Maine, filed by the EPA in its proof of a claim against Lincoln Paper & Tissue. A dreary recital of breakage and blunders, all resulting in pollution of the river and streams on the property, fills pages of court and agency documents. Fish kills, spills, oil sheens. Unlicensed discharges of untreated industrial wastewater, backed-up sewer lines, boiler water feed lines ruptured, overflow of the paper mill sewer collection box, untreated process wastewater and cooling water discharged into a stream on the property, civil claims, consent decrees, fines. The company agreed to pay a $1 million fine in 1990, at the time the largest ever assessed in the state, for violations of air and water quality and solid waste laws stemming from those discharges.

The State of Maine, beginning in 1987, issued advisories warning that consuming fish from the Penobscot River posed a serious health threat because of dangerous levels of dioxin in fish. These warnings persist today. Dioxin is a potent and persistent pollutant that causes cancer in multiple species, including humans, disrupts the endocrine system, and suppresses the immune system. Dioxins form during pulp and paper manufacturing when chlorine-based chemicals are used for bleaching, and those dioxins are released into rivers via discharge of wastewater. The company did not eliminate chlorine bleach in its processing until 1999. Problematic as pollution remains today, Kusnierz well remembers when it was even worse. When he first started working for the tribe thirty years ago, Kusnierz remembered the river as a dying, sick place, and families were fearful of going anywhere near it. "That was really, really sad. You couldn't see through it," he said. "The river itself stank. There was foam, the water was dark, it was impossible to see the bottom." There was so much heat in discharges from the mill, there were places that even in Maine winters, it didn't freeze.

In July 1999 an inspector from the US Bureau of Indian Affairs

(BIA), Jeffery Loman, visited the Penobscot Indian Reservation and took a trip on the river in a small boat. In a letter to former US Senator Olympia J. Snowe, the director of the Office of Trust Responsibilities for the BIA, Terrance L. Virden, went on at length as to what the inspector found, a letter rare in its evident fury. The inspector's first observation was a sign warning fishermen of dioxin pollution in the river. While traveling upriver, Loman became aware of an acute odor, ever increasing as he approached the outfall where Lincoln Pulp & Paper (as it was then named) discharged into the river. At the discharge pipe, Loman saw large clots of foam, and upon clearing the foam at the pipe, he found a dark brown substance that appeared to be much like rotting grapefruit spewing out in great quantities—a foul material that appeared to be heavier than water. Loman sampled surface water flowing from the mill property and tested the pH of streams and found it to be highly corrosive. Along the bank of the stream he sampled, Loman saw a dead bird, possibly a woodcock. Virden wrote in his letter to the senator: "My general impression of the Penobscot River, based on the observations Mr. Loman reported during this trip are simple: it stinks, it makes you sick, you can't eat the fish, and it's killing the birds. Clearly, it's not the way it should be."

The tribe's offers to improve the situation over the years were not welcomed—not by the company and not by the town. "When we were trying to get things cleaned up, the town was not supportive at all. It was, 'You are messing with our primary economic development,'" Kusnierz said. "We went up there and said, 'We are not after your jobs, we just want a clean environment, which you guys should want too.' It was not well received." Over the ten-year period in which mill after mill has shut down on the Penobscot River, water quality has improved, Kusnierz said. "Dioxin levels are coming down. But I'm worried people are completely turning away from the river because of their fear. Then there is the other side: you tell some people it's not safe to eat the fish, but they do it anyway. That's part of the culture." He remembers when he first got the results back from EPA laboratory tests on sea-run fish coming back to the Penobscot after removal of the dams. "We were restoring fish populations, but they were contaminated," Kusnierz said. "It's like, 'Oh this is great news, we got the test results back.' I got really emotional, my voice started quavering. How do I tell them they can't eat these fish? We brought them back, but you can't eat them."

In 2020 the tribe reported cancer rates among the Penobscot people are more than double the state and federal rates. "We have a situation where tribal members are literally dying by practicing their own cultural traditions their ancestors have done for thousands of years," said John Banks, the former natural resources director for the tribe from 1980 until his retirement in 2021. "It is affecting not only our health but our cultural integrity because the river is no longer available as it was in the condition it had been for generations upon generations." The Penobscot are among the Wabanaki, the Dawnlanders or People of the Dawn, the Algonquian Indian tribes of the sub–Saint Lawrence River and northern New England area called the Dawnland. Their ancestors include the Abenaki, Penobscot, Passamaquoddy, Maliseet, and Micmac, among other Native peoples. Their homeland has been inhabited since the end of the last ice age some twelve thousand years ago, the time of mastodon, mammoth, musk ox, and large herds of caribou, wrote Darren Ranco and Barbara Harper, coauthors of the 2009 report *Wabanaki Traditional Cultural Lifeways Exposure Scenario*. The many lakes, wetlands, and rivers running to the sea were the gifts of the retreating glaciers.

This, Ranco and Harper remind us, is where the Penobscot people have always been. They have never been removed, never federated with another Native group. And Ranco and Harper report this beautiful fact: the plants and animals the Penobscot people use today are the descendants of the plants and animals their ancestors used. "They share kin with one another, and all beings, back to their creation story from the brown ash tree," Ranco and Harper wrote. Survival for the Wabanaki meant deeply knowing this land and these waters in all seasons. What was where, when it was there, in what quantities, whether it was food, medicine, or material for clothing, for building, for boat building, for tools. The land was alive with the same foods the tribe values today, from white-tailed deer to moose and black bear. And the many other four-footed ones—the beaver, fox, bobcat, fisher, marten, otter, skunk, muskrat, porcupine—all with their use and part of the woodland circle of life, Ranco and Harper wrote. Other animals now are rare, such as the lynx. Or gone: caribou are no longer found in Maine because its old-growth habitat is gone but for rare patches. With its berries, seeds, roots, tubers, tree nuts, bark, maple sap, and medicinal plants, the forest provided all the Wabanaki people needed. The rivers, streams, and ponds were alive with fish, and the tidelands squirting with clams and studded with oysters. Lobsters, seals, porpoises and whales, and a multitude of sea ducks and geese completed a seasonal round, Ranco and Harper reported.

European colonization began as early as 1607 in this area. Diseases brought by traders and explorers and settlers, to which the Native people had no resistance, killed as many as 75 percent of the Wabanaki people in Maine, according Ranco and Harper. When Europeans arrived, some encountered already plowed and cleared fields, with no inhabitants—and few Wabanaki to repel and resist the European invasion of their territory. The Europeans, arriving from their cutover and thickly-settled land, were astounded at what they encountered. Here was a vast watershed cloaked with virgin forest to the waterline and rivers surging with raw power. The Penobscot River drains an area of 8,570 square miles, with a total fall of 1,602 feet from its headwaters, winding through steep mountains including Mount Katahdin, the highest point at 5,269 feet, and extensive wetlands, marshes, bogs, and wooded swamps. Translations of the Native place-names reveal the abundance that was here: Clam Place. Abundance of Eels.

The Wabanaki utilized the entire watershed, with the river providing the means for transportation. Families migrated seasonally to the coast in summer to eat lobsters and clams, and get away from the mosquitos, returning in the fall to family hunting grounds and camps, John Banks told me. "They would stay out of each other's territory, and that goes on to this day." Theirs was an affluent society. The abundance of their lands and waters, carefully managed with cultural protocols, allowed time for socializing, education, ceremony, leisure, oratory, and art. Still today for the Penobscot the river is a relative, and not just metaphorically. The tribe's five-thousand-acre reservation includes two hundred islands in the river, including Indian Island, the south-ernmost—a small island of about three hundred acres connected to the mainland by a bridge that is home to the tribe's main settlement. The tribe is small, with about 2,398 members. Many people left the reservation because there was not enough work to stay. But they have never forgotten the river. "We are not like some corporate entity that will shut down and move when there is no longer a profit to be made. We will be here forever," Banks said. "We don't have shareholders, we have citizens and the vast majority of them live here." The tribe in 2019 adopted a resolution enrolling the Penobscot River as a tribal citizen. That relationship endures despite all the people and the river have endured.

This resilience began with the Penobscot's ancestors, who adapted, survived, and persisted amid the coming of the explorers, traders, and settler-colonizers. Native people were considered highly-valued guides in the woods and on the rivers, and they appear in many ex-plorers' narratives, including Henry David Thoreau's essays on the Maine woods. As newcomers pushed in from the woods and upriver to cut the forest, Native people trapped the woods and rivers for traders. They logged and ran the rivers in the spring drives. Banks's great-grandfather, Lewis Ketchum, was a guide on the west branch of the Penobscot River and a foreman on the west branch log drive—one of the most formidable jobs in the woods. It was not unusual for men to be drowned on these drives, their bodies if they were ever found buried in the woods, and their boots hung in the trees. According to Banks, when Ketchum became sick while on the drive, he went

Penobscot elder and Penobscot language teacher Carol
Dana keeps this photo of her ancestor, nicknamed
Molasses Molly, by the front door to guard the house.

into the woods and wrapped himself in his moosehide blanket for
two days, to recover. Then he ran for two days downriver to catch up
with the drive. "When he walked into camp, they thought they were
seeing his ghost," said Banks.

Water-quality standards set by federal and state agencies don't take
into account the amount of traditional foods that Native peoples historically ate—and would like to eat again. Eating a modern American diet not only denies the cultural practices that are part of tribal
identity but also contributes to diabetes and obesity in people whose
bodies for thousands of years were nourished on First Foods such as
fish, roots, deer, and berries—not white flour and sugar and fast food.
Atlantic salmon were always part of their diet. But fishing is prohib-
ited for the King of Fish today, so scarce are they. Atlantic salmon are
federally protected under the US Endangered Species Act, beginning

with a population in the Gulf of Maine in 2000. An expanded listing included the Penobscot, Kennebec, and Androscoggin Rivers in 2009. Other fish, while not listed, are too toxic to eat. "Shad, lamprey, bass, alewife, we don't recommend it," said Jason Mitchell, a Penobscot tribal member and water resources nonpoint source coordinator for the tribe. "We have all these fisheries rights, but we can't utilize them because of the pollution." Traditionally people would have gathered alewives to fertilize the soil in their gardens. "But we don't dare do that either," Mitchell said. "I hope things keep going in a positive direction, that we can enjoy these fish that are a large part of our culture that sustained us for thousands of years. That we can eat these fish again."

The US Department of Health and Human Services Agency for Toxic Substances and Disease Registry, in its report on a health consultation for the Penobscot Nation released in May 2021, found tribal members who ate traditional foods from fish to turtles were at risk because of exposure to contaminants at levels of health concern, including increased lifetime cancer risk from PCBs, dioxin, and methylmercury. The agency recommended children younger than eight years and women who are breastfeeding or might become pregnant eat no Penobscot River fish. Dioxin was found in fish tissues at levels high enough to be a health hazard for Penobscot Nation tribal members of any age, including significantly increased risk of liver cancer. Wildlife are also contaminated. In their paper published in *Science of the Total Environment*, published in March 2021, Kusnierz, the water-quality specialist for the tribe, and his coauthors found mercury in the tissue of six species of migratory fish—alewife, American shad, blueback herring, rainbow smelt, striped bass, and sea lamprey as well as the roe of American shad. Levels were high enough to also pose risk to mink, otter, and eagle.

For the Penobscot people the river is their life—and that is not just a saying. "We've had to settle on processed foods. There is a lot of diabetes and obesity on the island," said Charles Loring Jr., who took over from Banks as director of natural resources for the tribe.

Penobscot elder Carol Dana wove this basket from the bark of an ash tree. The invasive emerald ash borer is killing off ash trees, imperiling ash bark basketmaking, one of the oldest art forms in New England.

"Substance abuse, people not being able to connect with our relatives, it's not going to change overnight. That is something I've seen across the tribal community. How could you miss it, you wonder why do we have the issues like we do, but when you put it all together, it's obvious. This is about more than fish. It's our whole existence."

Mitchell's great-great-great-grandfather, Big Sabattus Mitchell, was among the most famous of the Penobscot Indians who were top river drivers during the 1870s. He was the first to run the falls solo in a bateau on the Sourdnahunk Falls on the west branch of the Penobscot and survive. "They were log drivers. So it's very natural for me to be working on the river," Mitchell said. He knows what he wishes for his generation and those to come: "That someday the Penobscot people would be able to eat fish again and not be poisoned. That there is recognition that we are river people. We were stewards of this

Pollution caused by paper and pulp mills has left a lasting legacy of toxic pollution in Maine rivers, including the Penobscot, from which the Penobscot Nation warns its members not to eat fish. In this June 1973 photo, effluent from the International Paper Company Mill at Jay, Maine, gushes into Allen Brook near its confluence with the Androscoggin River. The mill changed hands multiple times before shutting down in 2023, following an explosion. Photo by Charles Steinhacker, National Archives.

river and we lived with it for twelve thousand years. The river is our relative, a citizen, and our nation is all the fish, and all the animals. They are all our relations."

After that day prowling around the abandoned mill in Lincoln, I went back to talk to town officials about what had been left on their hands since the paper mill suffered a catastrophic explosion in a key piece of equipment and then went bankrupt. Bronson, the town manager, said it pains him society didn't think about what would happen when this town's mainstay was no longer a paper mill. "You could say they just abandoned us. They are not here to clean this up." The final report on the site and its condition filed by the EPA in 2019 found it will cost about $60 million to remediate the contaminants at the mill, take down the buildings, and remove the debris. With about five thousand residents, the town does not have the tax base to manage such a burden, Bronson said. So it has been tackling the problem gradually, using federal grants that typically have only two-year cycles. "We just keep plugging away. It's going to be a long process." Lincoln has acquired the site and divided it into seven nonprofit corporations, to allow the town to apply for more grants in a year. Cleanup of the heaviest contamination was done first, including a lot of asbestos. Not that it was a smooth process: long remembered will be the day the feds showed up to deal with the toxic liquids at the site—but set fire accidentally to the tanks, creating a hazardous-materials incident. "We all know that saying, 'I'm from the government and I'm here to help you," Bronson said with an eyeroll.

One of the first paper mills built on the Penobscot, the mill at its peak of production employed some four hundred workers making good money, including retirement and health benefits. It was not unusual, said Lincoln economic development administrator Ruth Birtz, for three generations of one family to work at the mill, with college students coming back to work during the summer. Birtz grew up with the mill, so she understands as only a resident would the tolerance for pollution that meant paychecks to put food on the table. "When I grew up, it was common to have periods of time when you got covered with smog, and it had this stench and smell, and people would say, 'That is the smell of money.' That was our employment and we were very aware it was our primary source," Birtz said. "It was, 'Yeah the smell is bad, but what it gives us in the end, we are willing to tolerate that.'" When the mill faltered economically, residents pitched in to get it back up and running.

With the exit of the paper industry, the town of Lincoln is trying to find its new future. When the mill closed, the town council knew it was coming, and was already trying to diversify, Birtz said. But it was still a devastating blow. The school lost enrollment as people moved away. The hospital went through bankruptcy. "The paying customers they had were the mill employees, that only left them with the non-paying ones," Bronson said. The town had to regroup, get beyond the name Stinkin' Lincoln that went with its tie to this mill. "The town government was of a mindset that this was coming, they had taken steps to rebrand Lincoln and the town it could be. That was not that smell that came with the town," Bronson said. "That was to diversify the businesses, and not just the primary one that was here."

WITH ITS MANY LAKES and a seasonal tourist industry, today the town of Lincoln is pushing a fresh image and trying to back it up with new industry. The latest hope is in a lab at the shuttered ND Paper's pulp mill in Old Town. Lincoln signed a lease in 2022 for a small-scale start of a sawdust-to-fuel products venture developed at the University of Maine's Forest Bioproducts Research Institute, headquartered in one of the buildings at the Old Town mill. What if, said lab director Hemant Pendse, new projects for cellulosic fiber could be developed? Cellulosic fiber is the material in wood used for pulp and paper—but Pendse said it also has potential for other industries being hatched at the lab. He showed me through the lab, a maze of pipes and tanks and motors, condensers, agitators, reactors and controls, dials and gauges, straight out of a mad scientist movie set. In a corner of the lab were stacks of dried, pressed pulp that after the mill shut down, never made it on the boat to China to become boxes. The work here now was investigating at the bench scale new products from cellulosic fiber. Sawdust, old carboard boxes, logging residue including wood pulp. Any woody biomass at all can be converted into crude oil with heat and pressure, Pendse said.

So what might be possible? Lincoln had just committed to a twenty-year lease to Biofine Developments Northeast Inc. for a portion of the property at its former mill site—now called the Lincoln Technology Park—to develop a biorefinery plant, using the process demonstrated at the Bioproducts Research Institute lab. New life, from one dead mill to another. Town officials are cautious in their optimism about any of this. They have seen plenty of headlines in the Maine papers with photos of well-dressed people getting out of private planes making promises about new ventures planned in mill towns like theirs, Bronson said. There was the Bitcoin firm, the aquaculture firm, the biomass plant to produce bio carbon to absorb pollutants. "They were better than a tire kicker, but they were never going to hatch their egg," Bronson said. "There have been a lot of tire kickers. We have to sort our way through them." The sawdust-to-biofuel start-up still needed to get a dozen permits and financing in order. But the town felt sure enough this time to put out a press release about the lease. The difficult thing was managing expectations, Birtz said. "Probably one of the hardest things people have understanding, is people think it is going to happen right away."

The only people more eager for that new start than this northern Maine former mill town might be the people who cut trees. They are losing markets as paper mills shut down, so they too are looking for new products and possibilities for their work in the woods. I was surprised, but shouldn't have been, when I met the Professional Logging Contractors of Maine's 2023 loggers of the year, Molly and Alex London, and learned they had visited Lincoln just the day before me.

8

NEW BEGINNINGS

THE TREE CANOPY TREMBLED as Arthur Hathorn started up the machine and began rumbling into the woods, flattening a trail as he went. These were tiny trees, just saplings and poles, with a few larger trees, grown from a clear-cut in 1992. The goal on this job was to thin out small trees and leave the larger ones to grow, to improve the quality of the remaining stand over time. This was formerly paper company forestland, off the Stud Mill Road in Downeast Maine, and these thinnings were now grown to barely merchantable size for pulp. Hathorn was working to a set of instructions, in a prescription from the forester for the management company for these lands, privately owned by one of the world's largest landowners. He was to cut 40 to 60 percent of the stand, taking first any fir six inches in diameter and larger, any damaged trees, and aspen and white birch eight inches and larger. He was to leave any undamaged spruce six inches or bigger around, and yellow birch, particularly the strongest, most wind-resistant trees.

This thinning work is where loggers like Hathorn, on equipment like this, a cut-to-length processor, show their skill. Hathorn scrolled and tapped his fingers on the screen of the iPad suction-cupped to the inside of the cab of the wood-processing machine. The lines he drew on the screen he would now cut into the forest. He hit the accelerator and got rolling. His task was to cut the trees as directed, which required knowing their species on the fly and their value, stacking them into like piles, and above all, to not make a mess. The trails from his giant machine shouldn't eat up more than 20 percent of the harvest area—and he tries to keep it to 15 percent. He was supposed to keep

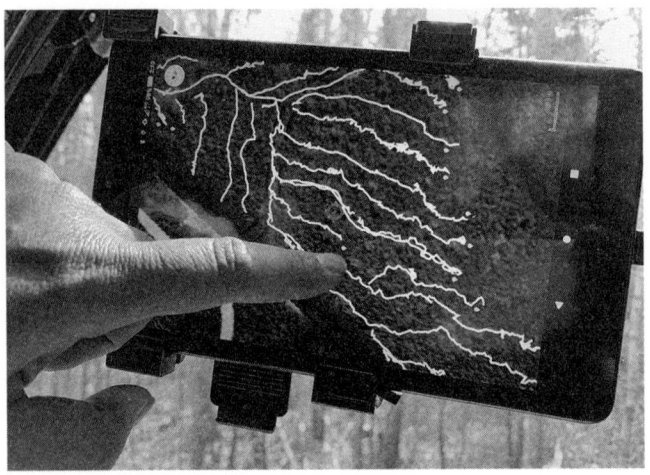

Logging today involves expensive, sophisticated equipment.
Logger Arthur Hathorn can cut, strip, and stack a log in less than a
minute, working from a prescription beamed to an iPad suction-
cupped to the window of a cut-to-length processor.

any culverts and drainage areas intact, put down brush in the trails
to prevent erosion, and leave the cut area in condition to regenerate.
His employers, Molly and Alex London, are thinning contractors with
crews and equipment that specialize in that service. The Londons
chose this type of machine in part because it rolls on rubber tires with
chains, instead of on steel tracks (like a bulldozer) that tend to have
more of an impact on the groundwater and soil compaction.

"It's hard to stay perfect," said Hathorn, who prides himself on following the prescription beamed to his iPad with careful woods work.
"But they don't come out with a tape measure, the only time they do
that is to measure the wood." The air-conditioned, self-leveling cab
on this processor, with its comfortable seat and joystick operation,
transformed logging for him. It's a job even a computer gamer could
love, he joked, with the electronics, the comfy seat, and joystick controls. As he slowly went along, Hathorn eyeballed the trees, looking
for the ones that fit his prescription for the cut. With the push of a
joystick, he reached out the machine's thirty-two-foot mechanical

arm to the base of a small tree to the side of the trail. He pushed a button on the end of the joystick and with a squirt of sawdust, the trunk was sliced through at the base. He grabbed it with the grapple on the same cutter head, swung the arm to the side of the trail he had just made, and zipped the head up and down the trunk, whizzing off the branches. He lay the stripped log on the ground, and reached into the forest to cut another.

Squished into the cab to be along for the ride, I timed Hathorn: just twenty-five seconds, from living tree to log, ready to be picked up and taken to a mill. At eleven inches in diameter, this was the biggest tree I watched him cut, and one of the few trees from this cut that would be big enough to create a sawlog for two-by-four stud wood, a higher-value wood product. "That's too small, I won't take it," Hathorn said bypassing spindly trunks. "The glory days, we don't see them no more," he said, as he worked a patch of scrawny trees. "That can go to the mill and make COVID swabs," he said, grabbing a tree so small it broke. "Can't win them all," he said and kept on rolling. "We used to do pine and hemlock for pulp, now the only reason I cut it is if it's in my way. That's a cedar," he said. "I can't get anything out of that." He let it be. "This will be two-by-fours," he said, and grappled a spruce. "That's a big one. Be nice if we were in woods like that all day." Another tree's sawdust squirted out the bottom of the cutter. He kept up a steady pace, one gush of sawdust after another, three more logs cut, stripped, and stacked in less than a minute, as Hathorn worked with the precision of a surgeon in this, his forest operating room.

He knows logging is controversial, but over his career Hathorn said he has watched the regulations he follows change to be more protective. "We used to run through mud holes, drive right through brooks to get wood on the other side, we would take it all. Oh, now we got to go through a bunch of crap," he said good-naturedly. "Everyone has his own opinion, a lot of people don't want it cut. They must have to use toilet paper." He had cut three 856-foot-long trails into the forest that day. Up at 3:00 a.m. to make breakfast and lunch then drive two hours to work, it was time to drive the two hours back home, and get ready to do it all over again the next day. The machine had recorded his work, and he would send the record of the tracks he cut to the home office. Hathorn backed the processor out of the trail he had cut easily as if on a driveway, parked it in a clearing by the logging road, and climbed down to meet the rest of the crew working in these woods today, and his bosses, Alex and Molly London.

The Londons talked about how much they love this work, this crew, and these woods as we all swatted bloodthirsty black flies. "I wish more people understood the skill level of our employees," Molly London said, bouncing a baby on her hip, the couple's newest of their four children. She is the first woman to ever serve on the board

Molly and Alex London—with their newest baby—on the job along Stud Mill Road in Downeast Maine, in May of 2023.

of the Professional Logging Contractors of Maine and proud of the people who work these woods. "There is this misperception, 'Oh just some dumb logger.' They are incredibly skilled, hard-working, they might not have a high school degree, but they have a huge amount of lived experience." Their company pays health insurance, provides a pickup for transportation, and pays full rate for travel time. The pay is good—from $45,000 to $65,000 a year—and the Londons provide a 401(k) program with a 3 percent match. "The ones who know college is not an option for them, they don't know about it. But this is good work," Alex London said.

For him, running this small logging business was a way to spend time in the woods with his family. "I have spent more time outside with my kids looking at the woods than a lot of people have in their whole life," Alex said. "I like the challenge of finding ways to get every bit of production out of these machines, trying to get the price you need to survive every day." To him these woods are a safe space, where he is most comfortable. Ron Martin, operating the machinery to take the wood Hathorn cut out to the road, agreed. A forty-year veteran of working in the woods, Martin wouldn't have it any other way. "I'm basically all by myself all the time, I don't have to deal with people. I love the woods. I can't work in public, I get nervous, I don't like being around a lot of people."

Jamie Brackett reached out a meaty hand, introducing himself. Brackett also runs a processor, cutting trees—just like his ancestors did, selecting and harvesting trees, only under much different conditions. Brackett has worked in the Maine woods since he was sixteen. "It's a good job, the paycheck, I guess, and the freedom of it, I like that," he said. At forty-two, he had survived decades of work with a chainsaw with a twenty-inch bar that weighed thirteen pounds. "Put a saw through my foot, dislocated my shoulders and knees, got knocked out . . . " I had the feeling he could go on, but he stopped. "It's just what my family always done, my father, my grandfather, he worked with horses, on the log drives on the Kennebec. It's not a thoughtless job, we are not just Neanderthals out here, a lot more goes into this job than they think," Brackett said of the public. "The wood industry is not what it looks like to other people, we are not just randomly hacking stuff down." He waved off the black flies and hopped in his truck to escape. "Don't forget when you go to the bathroom," he yelled back at me and grinned.

The Londons are the fresh face of an industry that is aging. One reason is that it is a hard industry to enter these days, with the enormous cost of the big equipment used for working in these woods. It took a stake from Alex's dad to buy the sophisticated and fuel-efficient equipment that does the quality of work they want to be known for, Molly said in our drive together to the worksite that day. Today a new processor such as theirs runs about $625,000, a forwarder another $450,000 for bringing the logs out of the woods, and a logging truck about $250,000 to $275,000. The Londons started with all used equipment, eventually building enough equity to get their first brand-new forwarder. Certified loggers, they started their company in 2017 and have never looked back. "It was, I want to do this, and I am not going to be scared—we are just going to do it," Molly said. "The profit margin across the board in the logging industry is like 2 to 3 percent, and that was before COVID and crazy inflation. I can't even imagine what it is today. But we are here. We are doing it." For the Londons the challenge of running their own business makes it worth it. It's a way of life; it is not just a job.

Their company works for a range of owners, and they cut wood for a variety of products, everything from chips for pulp to stud wood. Forestry in Maine is no one thing, and the cutting reflects the goals and values of the various landowners. The Londons' biggest client is the Appalachian Mountain Club (AMC). This is forestry with a lighter harvest, with a focus on leaving larger trees to grow, while still cutting wood for revenue, and to open up even-age, former plantation lands for regeneration. Other protected areas are not harvested. "We have learned a lot from the way AMC does it, you can be conservationists,

be concerned about the environment and still harvest wood," Molly
said. "You can have multiple uses, retention areas with old trees, and
keep the forest working. We think as the world moves on, more and
more landowners might want to see that, instead of just locking up
the woods."

The challenge for people like the Londons and the loggers they
employ is that consumer tastes and markets are changing. The work
they do in the woods is sensitive, within hours, to the global market
cycles these forests are part of. As he worked that day, Hathorn was
skipping small pine because there was no market for it, since the un-
expected closure just weeks earlier of the Chinese-owned pulp mill
in Old Town. Suddenly low-value wood had no value. "I couldn't
believe how fast it happened, within hours of the announcement
my phone was ringing, it was 'Don't cut anymore, and get every stick
out,'" said Alex. "I couldn't believe how much we were impacted,
and so quickly." Fewer markets for low-value trees normally cut for
pulp is bad news for people doing the cutting. That's what took the
Londons to Lincoln, Maine, just the day before I had visited with
the town managers dealing with the dead mill in *their* town. For the
Londons the proposed biofuel start-up was interesting because if it
goes anywhere, it might help keep their loggers working. There's not
much margin for a downturn.

"It's not a very profitable industry to be in, there are a lot of hands
in the pot before it gets to the logger, it is very similar to farmers,"
Molly said. But they love what they do, where they get to live, in the
least populous county in the state of Maine, and running a small
farm too. "It's crazy chaos all the time. Very happy chaos," Molly
said. She wants to see these lands stay forested—and people working in these woods into the next generation and beyond. "I believe
it is really important to look at the jobs we have created in our town,
the families that we support, a lot of our guys are the sole income
for their families," she said. "That is really special to me, these are
really skilled, hard-working guys. I feel forest management is really

important for a lot of big-picture reasons. I would love for it [the North Maine Woods] to stay working forest. That is what it has been forever." That day, each of the crew reminded me their work is where so many of the products come from that people use every day. Toilet paper came up three times. And it's true: in forests like this, and in the work of people in jobs like these, is where some of the industrial paper products from low-value wood—everything from newspaper to paper bags, toilet paper, all those cardboard boxes from Amazon, much of what people use—come from. The scrap from higher-value cuts also gets made into chips and pulp.

FORESTRY HERE in most instances looks very different from the practices in the Pacific Northwest and British Columbia. The logging roads cut in to mountainsides, the clear-cuts that reach over the horizon, the herbicides and replanting with a monoculture—there are some large operators who do that here. But mostly, this was not like the industrial plantation lands I was used to in the Pacific Northwest, with Douglas fir cut on forty-year rotations, sprayed and replanted for the next round of cutting, possibly on an even shorter rotation than the last one. Mostly, foresters in Maine cut the forest, walk away, and let the forest naturally grow back. Cut hard for centuries, not for nothing do foresters call the North Maine Woods the magic forest. With plenty of moisture that comes evenly through the year, lots of flat land, and a diversity of tree species, logging here is more like mowing than logging. The result is a young, small forest of surprising species diversity. From the road it looks almost suburban and surprisingly uniform, a wall of low trees close to the roadways that stretches unbroken for mile after mile.

It's impossible to know what's back in there beyond the beauty strips left by the loggers along the road, so solid is the wall of trees. I wanted to get in there, to see the north woods I had heard so much about, and Molly had a suggestion: a tour with Steve Tatko, of the Ap-

Maine's northern forest has been cut and recut so many times, today the harvest from most stands is of small logs, like these.

palachian Mountain Club. These were the forests the Londons' crews were working in, and where Alex's father, Bill London, was working with the club to take out culverts on the privately-owned logging roads he had helped build, lacing through these vast timberlands. Tatko, vice president of land, research, and trails for the AMC, arranged a stay for me in one of the club's wilderness lodges, a collection of cabins arrayed around a main kitchen and lodge in the north woods.

These lodges are a tradition Mainers have always cherished in these woods and a featured experience offered by the Maine chapter of the AMC. The oldest outdoor club in the United States, formed in 1876 to explore and preserve the White Mountains of New Hampshire, the club has expanded throughout the northeastern United States with twelve chapters from Maine to Minnesota. The AMC has a conservation mission in addition to recreation, and in Maine that was even taking the form of buying up land to ensure public access. Its Maine Woods Initiative is a multiuse land and conservation project in the famed 100-Mile Wilderness, in the heart of the northern forest. Since

2003, the club has permanently protected more than 114,000 acres of this forestland, with more acquisitions planned. It's a big vision, including creation of 130 miles of recreational trails, three off-grid lodges, a responsible forestry operation, and even restoration of watersheds for native brook trout and endangered Atlantic salmon.

Tatko fired up his truck and we headed out past a signpost riddled with bullet holes: this was the Golden Road, the privately-owned, unpaved nearly hundred-mile-long highway through the woods once roaring with logging trucks hauling pulpwood to the Great Northern Paper Company's now dead paper mill at Millinocket. We bounced over the man-eating potholes, Tatko narrating the landscape as we went, as the forest, young and short, unspooled by the mile out the windows. "You are driving through a forest that capitalism has pushed to this condition," Tatko said. "We are cutting down trees that are as old as we are, and they can live for centuries." I had asked to see the Ripogenus Dam, about thirty miles west of Millinocket, still strangling the lower west branch of the Penobscot, even though the mills it powered for Great Northern Paper are long dead.

Today the dam's electricity is sold by the Canadian power company that owns it. This dam in the middle of nowhere, owned by nobody from here, felt like one big hostage taking in a landscape and a river long controlled by outsiders. As we took in the dam's grim face, Tatko explained that outsider control has been tolerated by Mainers for generations, in part because of the long tradition of public access even to private land. They might not own much, or run anything, but even the poorest families could still use the privately-owned forestlands all around them for hunting, fishing, and camping. This tradition was crucial, for unlike the Pacific Northwest with its millions of acres of national forests and parks, in Maine today there is still almost no public land. About 95 percent of the state's forestland is privately owned.

But the old dynamic was shifting, with the sell-off of much of this forestland in the 1980s and 1990s to distant investors, managing a financial investment that just happened to be a forest. Pulp mill owners used to own these forests to provide the trees for their mills, powered by their dams, to process the trees cut by their loggers, hauled on their trucks, over their roads. All that vertical integration blew apart with changes in the tax laws that gave forestland owners incentive to sell off their lands. At the same time, global pressure to produce forest products more cheaply was goading the US forest products industry to invest in wood productions overseas, chasing lower taxes and less regulation. While these trends were underway everywhere in the United States, nowhere was the change as dramatic as in Maine's northern forest, with more than 40 percent of the land changing hands from 1998 to 2016, some of it more than once.

"When I was born, there were seven landowners in Maine and they were a mix of family and corporate owners, but people knew who they were, and the people who worked for them," Tatko said. "There was that nineteenth-century sense of corporate responsibility, paternalistic ways of taking care of the community." The corporate raider era that started with changes in the tax laws and international markets kept going through the early 2000s. Local people were whipsawed by the changes. Willimantic, Maine, population 134 in the 2020 census, is forty-eight square miles of scenic forested beauty, and it's where Tatko grew up hunting and fishing and working in the woods. He was born in 1987, and the land has been bought and sold four times just in his lifetime. Such turnover upset what people long took for granted, Tatko said. "The ethos of public access to private land has been a cornerstone of how people saw themselves fitting in on this landscape that they don't own or control at all. For communities like that, you no longer know if you are going to have access to those places that people for generations have had access to," he said. "It was the financialization of the 1990s, that is what killed this place. The incessant greed of the eighties and nineties of corporate America, to push for maximization of the return of the shareholders. At least you knew who you were dealing with, and what their intention was. Now we have no idea who owns it."

The sell-off did have the effect though, Tatko said, of opening conversations about the future of the landscape that would have gone nowhere a generation ago. The changes set the direction for his life: he went to work in conservation with the AMC to help buy up part of his backyard to keep it in forestland, with guaranteed public access. Tatko is leading regeneration forestry on the property. The idea is to open it up with cutting both for revenue and to leave other trees to grow on former paper company land that was clear-cut about thirty-five years ago and has regrown to an even-aged mess. "As foresters, we are always dealing with the results of past decisions," Tatko said. This is a first step in regeneration forestry needed on countless acres all over the northern forest of these former paper company lands, he said. "If you manage conservatively, reduce the harvest, allow the stocking to regrow, focus on growing quality rather than quantity, have a more high-quality product that feeds niche markets. We are one of the few people that grow old large hardwood saw logs, they need it for flooring and furniture." This is restoration forestry that also can keep loggers like the Londons working in the forests they love, to develop mature, multiage, higher-value forests. Today the work the Londons do for AMC is 80 percent of their business.

Now that I'd seen some of the north woods on the ground, I wanted to get up in the air to get a different perspective on this vast forestland. The North Maine Woods is still the largest contiguous undeveloped forestland in the nation. Primarily privately owned, still largely unsettled, this northern forest mostly has never been converted to any other use. It is forestland of global importance for forestry, carbon sequestration, birds, water quality, and wildlife. It remains even today a place apart. The Unorganized Territories of the North Maine Woods—more than half the 2,295-square-mile area of the entire state—has 429 townships but no local, incorporated municipal governments, and only about nine thousand full-time residents.

Not long after I visited with Tatko, I hired a plane for two days of flying over the north woods. Thomas Coleman's day job is working as the upper Kennebec regional forester for LandVest. But he is also a commercial pilot, registered Maine guide, and licensed forester—the perfect person to provide the overview I wanted. When I called his private charter company Wings, Woods, and Waters, Coleman said he was game. He grew up in the North Maine Woods, and like his father before him, he would live and work nowhere else. He was eager to show off a place he had known since he was a boy, and said he would soon be putting the floats on the plane for the season. A few weeks later on a perfect May morning, I met Coleman at the seaplane base in Jackman, Maine, where his sweet little four-seater plane bobbed, ready to take off for the first of two half days of flying. I'd asked for a tour to show me the good, the bad, and the ugly of typical forestry in the north woods. We buckled in, put on headsets so we could talk, and with a sputter and roar our little plane lifted off.

I had asked senior ecologist Jonathan Thompson at the Harvard Forest in Petersham, Massachusetts, to join me on the tour and help interpret what we were seeing. Thompson and a team of collaborators are deeply engaged in studying the northern forest and its future. Thompson earned his master's and doctorate degrees at Oregon State University, worked on fire crews in the Pacific Northwest, and is used to the look of industrial clear-cut forestry we had both seen so much of on the other side of the country. This was different. From up in the plane, the land looked like a rumpled green cloak. Hard used, to be sure, including some areas that looked trashed. But most of it looked better than we expected: green, growing, and covered with trees. Thompson was surprised at what he saw. "This looks better than my data has been telling me," he said, of the view out his window. Of course it is hard to tell the height of trees from above, he added. But still, this was not the bleak uniform moonscape we had expected from what we had heard and even seen from satellite imagery and the taste I got outside Wytopitlock, Maine, on a previous research trip.

The big-picture reality in these North Maine Woods was far more complex. It is no one thing, and can't be simplified as one enormous

Seen from the air, Maine's northern forest is a patchwork of different versions of the same thing: cut-over forestland in various stages of regrowth. Islands of older trees are a rare exception, primarily seen in parks.

mown-down mess. There were bizarre cuts, with trees left in square patches, from up here looking for all the world like dice rolled on a table. There were careful manicures with trees regenerating fast and vigorously. Bald clear-cuts that looked like anything I'd seen in the Pacific Northwest. But mostly, what we saw were the herringbone shapes of strip cuts over just about everything, in different stages of recovery. This is the dominant harvest method today in Maine. Strip cutting, or nonselective partial harvest, was implemented in 1991 in response to public outrage at gigantic clear-cuts made by industrial landowners, salvaging trees threatened by outbreaks of spruce budworm, a native forest insect with periodic outbreaks that causes extensive damage to spruce fir forests. While the new law did not outlaw clear-cutting, it imposed new requirements. The result is what some jokingly refer to as ABC forestry: anything but clear-cut.

Clear-cut harvesting dropped from 40 percent of the harvest acreage in 1990 to less than 5 percent in 2000. But the total annual acreage cut *doubled* during the same time period, as landowners increased rates of partial harvesting to cut the same volume of logs that had been harvested by clear-cutting, as Erin Simons-Legaard, Kasey Legaard, and their coauthors found in their analysis of long-term outcomes and trade-offs of forest policy management on forestland in Maine. That fragmentation of the landscape is bad for some species of wildlife and is leading to a conversion of the forest to more deciduous, sun-loving trees. Henry David Thoreau's "dark like a standing night" forest is giving way to forests rife with poplar and gray birch. That's bad for warblers that depend on spruce for food. Because it is so pervasive, nonselective partial harvesting is homogenizing the landscape. These were unintended consequences, the authors of that paper pointed out. There was no analysis of how the policy change that discouraged clear-cutting might affect the future forest condition or sustainability of Maine's forest. And it may have contributed to the widespread sell-off of forestlands, the authors wrote: between 1995 and 2005 more than 80 percent of the timberland in Maine changed ownership, often two or three times as the forest—previously in a few large ownerships—became parceled and owned primarily by financial investment firms less tied to sustainable timber production than the previous industrial owners.

The worst outcomes were dirty clear-cuts—cutting as much as possible while staying just above the Forest Practices Act requirements, stripping the property of many values while still remaining legal. Big equipment poorly managed leaves deep ruts, muddy trails, damaged and broken trees, hashed and bashed. I thought of the cuts I'd seen covering so much of the landscape around Wytopitlock, the ruts more than a foot deep and trails grown up in weeds. Erin Simons-Legaard, assistant research professor in forest landscape modeling at the University of Maine School of Forest Resources, says the change on the landscape has been immense—and she's not convinced that's a good thing. "I started with an anti-clear-cut mindset. But that area-to-volume trade-off that happens, we could cut half the area and get the same amount of wood, and maybe there is value in that, especially if that mature forest is allowed to just be."

The clear-cut reform was not the only thing that has had unintended consequences in these woods. In Maine more than $50 million in conservation easements funded by the US Congress are on lands their owners have cut just as hard as lands without the easements. In fact, the rate of harvest on some parcels is even higher on the lands in conservation easement than without, Thompson and his collaborators found in an analysis of the transactions. The easements did not appear to provide substantial ecological benefits, their work found. The protection of lands actually at risk of conversion to development also was negligible, because development pressure is primarily in the state's southern half, not in the northern forestlands, where the easements were funded. Population and development pressure just is not there in these isolated lands.

Thompson saw a squandered opportunity. The $50 million in conservation easements could have rewarded landowners for keeping their trees growing longer—or bought the land for preservation—but instead paid investors for cutting even harder. "We get nothing out of all that investment in corporate financialized easements," Thompson said. "These [easements] have prevented two acres of forest lost per year. And harvest has increased by three thousand acres per year. I consider this to be a total boondoggle, a waste of precious conservation easements. It's an abuse. It doesn't have to be this way. If we are going to protect lands with federal money, it ought to result in some change in land use on the ground. Otherwise, what are we paying for?" Buying an easement to keep the lands in forestry doesn't make sense in a place where there is no development pressure. "We need a better match between threats and protection," Thompson said. "The pushback is, the benefits won't be realized for a long time, there is no threat today, but there may be one hundred years from now. But what was your discount rate on what we paid now? I would rather use this money where the threat is now. I couldn't be more in favor of conservation easements. But this is an abuse of easements that threatens to take away that tool."

The potential *is* there in the vast forestlands of Maine for alternative futures that embrace both local wood production and improved carbon uptake, Thompson said. "There are massive opportunities for good forestry, this is a landscape that wants to grow trees. The stocking throughout Maine is a fraction of what it should be, it is cut on shorter and shorter rotations. Cutting to [the rate of] growth is sustainable, but it does not lead to long-term saw timber, it leads to more pulp." One thing that would help create markets for longer rotations—growing trees longer on the land for higher-value products, such as saw logs—would be more people paying attention to where their wood comes from, and how it is grown. That, and being willing to pay for a local, premium product, as many do now with food, Thompson said. "I wish more people got as excited about local wood as local food; I never have seen two-by-fours at the farmers market, and I wish I did."

Thompson and his collaborators' new research posits alternative futures for these Maine forestlands, vetted by community discussion, in a project funded by the National Science Foundation. The goal is to pose alternative scenarios for this northern forestland, working with maps generated by researchers in Thompson's lab at the Harvard Forest that show results of different assumptions about land uses and forestry practices. What are possible future scenarios for this unique forest? One thing he would not recommend is offsets, in which an industry purchases credits from forestland owners, to keep the landscape in trees, for sequestration of atmospheric carbon, in theory offsetting pollution the industry continues to put into the atmosphere. Multiple studies have found such schemes to be just that, with forestland purchased that wasn't going to be cut anyway, so the offset was not truly additional, in terms of benefit to the atmosphere in cleaning pollution. Or the forest does not perform as promised, in terms of growth.

Worse still, the pollution continues—not only an ecological problem but a moral one, as noted by Charles Canham, forest ecologist at the Cary Institute, in New York. Canham's groundbreaking analysis in 2021 showed there are deep and structural flaws in offset markets that routinely allow gross overestimation of the amount of benefit actually gained. "The bottom line is clear—a great deal of money will change hands, landowners with large tracts of forestland will be enriched, companies will claim grossly exaggerated quantities of carbon offsets, and brokers will collect large fees," Canham wrote. As for sequestration—forgoing harvest to keep growing trees for their ability to store carbon—the big question is so-called leakage. If the trees aren't cut in Maine for forest products, where will they be cut? What is almost never mentioned as a pathway to forest retention is *demand reduction*. Shrinking our collective footprint on our forests. This is not a popular pitch in a country in love with convenience and an economy built on consumption. And there is this: If there is less demand for wood products, less cutting, what about the loggers, the haulers, the workers whose jobs are tied up with these products, who want to live in rural areas and keep doing what they do? This is some of what the Thompson lab, with community discussion, is exploring.

I WENT UP AGAIN in Coleman's plane, this time with Charles Loring Jr., director of the Department of Natural Resources for the Penobscot Tribe, and Ben Stevens, who is Passamaquoddy and director of forestry for the Penobscot. Both men have young families and are in their early thirties. They represent a new generation of young tribal leaders who went to the same schools as their non-Indian colleagues in forestry management and production. But they also bring their culture and values to their forestry jobs. Maine is a small world, and the realm of local natural resource professionals who know and love these woods and grew up here is even smaller. Loring and Molly London both went to the University of Maine and had some of the same classes and professors. Stevens, an avid canoer and kayaker, recognized Coleman, a whitewater enthusiast, from paddling outings when Stevens first showed up at the plane. Coleman also knew the tribe as a client, having done surveys for the tribe of its lands in the past.

This was the first time Loring and Stevens had seen the forest-

Ownership makes the difference in management of forestlands. At Alder Stream, a forest owned and managed by the Penobscot Nation, Charles Loring Jr., natural resources director for the tribe, explains the forest is managed for multiple values. These include income for the tribe, jobs for tribal members, water-quality protection, and wildlife and cultural uses.

lands they manage from the air. I wondered what they would think, seeing all this land that was still their traditional territories but now controlled by so many others—as it had been since settler colonization. All that we would fly over were (and still are) tribal homelands. But today this land bears the scars of industrial extraction. Coleman swept his seaplane over the land, banking, turning, and narrating what we saw, tumbling through thin air deftly as a sea otter in water. He and his plane were one, zooming, turning, talking, diving, climbing, as he rode the wind. It was pure fun and a unique perspective on a landscape where it is impossible to see the forest for the trees without getting above it. From up high, the hard use of this landscape was revealed—but also its remarkable resilience. In the middle of the flight a large green island of trees stood out taller, darker, and unbroken in its tree cover: this was Big Reed Forest Reserve, at five thousand acres, the largest old-growth forest left in New England. Purchased by the Nature Conservancy in a series of deals concluded in 1990 for permanent preservation, this remnant of what used to be was a good reference for the surrounding landscape. The reserve loomed tall amid a landscape that, while mostly green and forested, was a patchwork of different styles of the same thing: cutting.

Coleman turned the plane again so Loring and Stevens could get a good look at Alder Stream, what we had come to see. This was predominately Penobscot Indian Territory over which the tribe regained control as a result of the 1980 Maine Indian Claims Settlement. One of the tribe's parcels of forest trust land, managed by the tribe for the tribe, in perpetuity, the canopy looked fluffy and lush. Roads were minimal. "Looks great, you are getting wonderful regeneration," Coleman said. Seeing it from the air confirmed what I had observed on the ground with Stevens and Loring when we drove it and hiked it together a few weeks earlier. Although they are cut, Penobscot forestlands are managed for multiple values: consistent income, jobs for tribal members who get hiring preference, water-quality protection, and enhancement and protection of the cultural integrity of the tribe.

Clear-cuts, where they do them, usually are limited to less than five acres, Loring said, a little bigger, maybe eight acres, if they are done primarily to create open areas for wildlife such as deer. The tribe does not use herbicide on their land, and all regeneration is natural.

But there is this, too. Seeing the tribe's forestlands with Loring and Stevens, from the air and especially on the ground, revealed something else: attachment. They weren't just managing these lands for clients or unknown investors. This was their home place, forever. This 23,535-acre parcel was the largest single holding of tribal trust land. A few weeks earlier, as we lifted the gate to enter the tribe's land, it was clear right away how well these two knew this landscape—and how carefully they tended it. As soon as he was on the other side of the gate, Loring tested the ground by hopping on it with both feet. He wanted to make sure that Maine's infamous mud season had abated enough that we could drive the road without damaging it. He had a shovel in the back of the pickup, and seeing ponding in the drainage swale, he and Stevens immediately set to work clearing the blockage in the culvert on the other side of the road, setting the drainage right again. We drove on, there was a cut they wanted to show me. But we kept stopping the truck. A contractor had done a harvest for them recently and they were checking the work.

Everything caught their attention—they were like homeowners coming back to the house after the departure of a guest. Was that buffer wide enough, or was that cut a little close to the stream? What was with that slide on that hillside? How had that happened? The cut was fine, the contractor left a lot of good stuff as instructed, but why wasn't it neater? A tiny shred of a wrapper caught Loring's eye in the tangle of brambles by the side of the road. He picked it up. The tribe directs its contractors to harvest from the road, so there were no trails cut into the woods. The machine's arm is used to reach in and select logs here and there. The logs are piled on the road to haul them out—new yards for storage are kept as small as possible. This was forestry as gardening; Loring was pointing out where the most recent cut had been taken, but it was hardly visible, even in early spring before the foliage was out.

The results were the sum of many decisions. Using temporary bridges instead of roads and culverts where they can, to leave the land as they found it. Leaving trees more than fifteen inches in diameter behind in selective cutting that regenerates the stand, allowing more room for the larger trees to grow. Managing the land to nurture a diversity of ages and species—a forest, in other words. "The goal is making a little bit of money, and enhancing what is here," Stevens said. We weren't the only ones to notice the difference from industrial paper lands. A bull moose, just coming into its antlers, peered out from the woods, its coat wet with rain. Loring saw it first, stopping the truck and picking up his binoculars for a look. When we saw a strutting tom turkey, Loring played a turkey call on his phone, trying

Ben Stevens, Passamaquoddy, forestry director for the Penobscot Nation, notes the tribe uses no herbicides on its land and that all forest regeneration is natural.

to get it to gobble. The red-tailed hawk, the coyote, the white-tailed deer—all the animals we encountered that day—Stevens and Loring noticed them all. This was not just a job, and not just a site visit—it was a visit with relatives, with all the pleasure of catching up on the news. What had happened since the last time they were here, what was anticipated in the coming season?

Loring pulled the big black truck into a clearing where the most recent harvest had been piled for loading. He popped down the tailgate and broke out a small camp stove and fry pan. It was time for Maine's signature red snapper hot dogs all around. Dyed a frightening crimson pink, they were just the thing before starting the long drive back to the natural resources building at the reservation. What was really the difference between this and industrial forestry was the goal of the owners, pushing for maximum yield biologically rather than financially, Stevens said. "To me it's value, how are you calculating that, quantifying that. This is what does not show on balance sheets. Water quality. Different species, more cover. Ovenbirds. Heterogeneity. We are trying to replicate what nature does." What struck me as we talked over the day was what, despite their different backgrounds, the Londons, their crew, and Stevens, Loring, and Tatko all had in common: the love of these lands and waters, and the desire to live and work forever in this place that means so much to them. While they use this place, they are working with the future in mind. It wasn't whether the land was used or not, but *how*. And sometimes, on this landscape, things were coming full circle.

Tatko pointed out to me that one of the most important contractors in the Appalachian Mountain Club's restoration work is Alex's father, Bill London. As part of the club's revitalization work in the northern forest, he and Bill were on a blitz campaign to tear out old culverts on the logging roads—many that Bill himself had built—and replace them with bridges so fish could more easily pass through underneath. *Sea-run* fish. This is something no one in these woods for a century had even thought to concern themselves with. "Salmon and sea-run fish is something no one ever thought they would see in our lifetime," Tatko said. "Seeing those fish coming back to interior Maine." Yet here was Bill London, at sixty-seven, using the skills he already had taught himself working in these woods to make way not only for logging trucks but sea-run fish. Suddenly old names like Silver Lake had new meaning, as shining silver river herring migrated home from the sea, reconnecting these forestlands and the sea for the first time in more than one hundred years. "We have half-a-million alewives in Silver Lake now and we have Atlantic salmon coming back," Tatko said. "Here is this globally significant forest that sea-run fish are starting to return to. It is all interconnected, people here get it. People feel it." He views the AMC working in these woods as a pivotal moment.

> People have to revalue this place, that is what the conservation movement has called Maine to do for the last thirty years, find value where you live, and find value in yourselves. It is really visceral. The change is basing management decisions about how we are going to manage this place where we are integrating fully the ecology and the needs of communities and using that as the initial lens through which we are making decisions. That is something that hasn't happened before, and we want to demonstrate it can be done. It requires a great amount of humility, you have to open yourself up to learning about the place, the community, the ecology, and the needs of the people. You demonstrate an interest by living there, showing up, hiring local people, giving local people the opportunity to make management decisions. All these people that have become part of the family, Molly and her contractors, that is the real magical element to this, being focused on people in the outdoors wherever they are at. It isn't about net present value, it's about tomorrow.

The goal now is to heal not only the forest and the waters but to reevaluate the long-held idea of outsiders making the decisions and setting the destiny for this land—and thinking the only choice for local people is putting up with that or leaving. "We were told overtly that there is no future here, that if you want to succeed you have to go to school and go away," Tatko said. "We are giving people a choice they didn't have before. But this only works if the places are healthy and the people are."

So far, in these northern Maine forestlands, the right answer is multiple uses on the land with an eye to the future, not just making a buck, and using the skills of local people to heal the scars of the past. To Tatko, the fish passage work is a great example, returning access for brook trout, with work by local people, with local materials, and local designs. "For a while, Billy and I were doing more culvert work than most states. People thought we were nuts," he said. "It can't be one big movement, it has to fit local geographies and people. But the values of self-expression, of enfranchisement in decision-making and a healthy economy and ecology, those have to be universal. We need to deconstruct those deeply embedded myths of common people not able to act in their own best interest. It's about home and a shared vision."

On another tour through these forestlands, I asked Bill London to show me the bridges he was making and installing all over these logging roads. He had by then replaced 106 culverts, on contract for the AMC with help from other federal, state, and nonprofit funders. It was a perfect use of the field engineering he had learned on the job, put to new use, so far improving access to ninety-six miles of spawning habitat for Atlantic salmon, alewives, and native brook trout. We stopped at one of his new bridges. London got out of the truck and scrabbled down the bank. I clambered down the hillside in the rain, wondering what he wanted to show me. Then I saw it: his name, written in welding bead. He is so proud of these bridges, he signs each one. "Look!" he said, pointing at the stream below, "that's a baby salmon!" I scooted down the bank and peered into the water, but of course the fish, whatever it was, was gone by then. Could that possibly be right? Here, in these woods? Not long ago, that was a question no one would have even asked or hoped to wonder about. But that's changing.

9

ABOARD THE
SILVER SMOLT

IN THE SPRING, when the river herring started running, Rory Saunders invited me to come along for a fish survey on the Penobscot River. So on a chilly, sunny morning in May, I met him at a dock where he was readying a humble craft: the *Silver Smolt*, an open aluminum skiff. Saunders, fisheries biologist for the National Oceanic and Atmospheric Administration (NOAA) based in Orono, Maine, and Justin Stevens of Maine Sea Grant conducted these boat surveys every two weeks, from earliest ice-out in spring to ice-in come winter, running transects all the way from freshwater in the river at Bangor to saltwater in Penobscot Bay. Their goal was to track ecosystem-level changes in the estuary, since the region had taken the once unthinkable step of dam removal on the lower Penobscot beginning in 2012. Saunders and Stevens had worked together on this survey since 2010, and the continuity of their observations is an invaluable record, of the river before and after dam removal as it comes back to life.

We bundled up, ready for fast travel. Saunders piloted us into the middle of the channel, and I held onto my hat and cinched the hood of my down jacket as he hit the gas. We skimmed along the river, the banks zipping by until we reached the head of tide at Bangor. Saunders cut the motor, and without a word he and Stevens got to work, heaving equipment overboard, to start the survey. A seal popped up his head to watch as the equipment, at four pings a second, began reading the water column with sound waves. This was an echosounder, basically a science-grade fish finder, used to detect life below the surface, from tiny plankton to shoals of fish. The goal was to generate an estimate of fish biomass in quarter-mile transects all along the

Rory Saunders (standing), fisheries biologist for the National Oceanic and
Atmospheric Administration in Orono, Maine, aboard the *Silver Smolt* for a spring
survey of the Penobscot River to assess biodiversity since dam removal.

estuary. They would be able to see the data in real time as the sonar
readings were displayed on a laptop strapped to the gunwale of the
boat. *Ping, ping, ping,* through thirty miles of estuary. The fish showed
up as dots on the screen, plankton as more of a smear.

Nobody really knows how to sample an estuary this size, and it
took years of experimentation to figure out what was safe, efficient,
and scientifically solid. But what an office in which to work: it was a
beautiful spring day, with a breeze pawing little wavelets on the river's
shine. Amid a day all silver and blue, a fizz of new green showed on
the trees alongshore. Saunders navigated the skiff, while Stevens dealt
with the machines and made handwritten notes on the birds and
other animals. He has closely observed spring on this river enough
times to anticipate the rhythm of its pageantry, the oaks and other

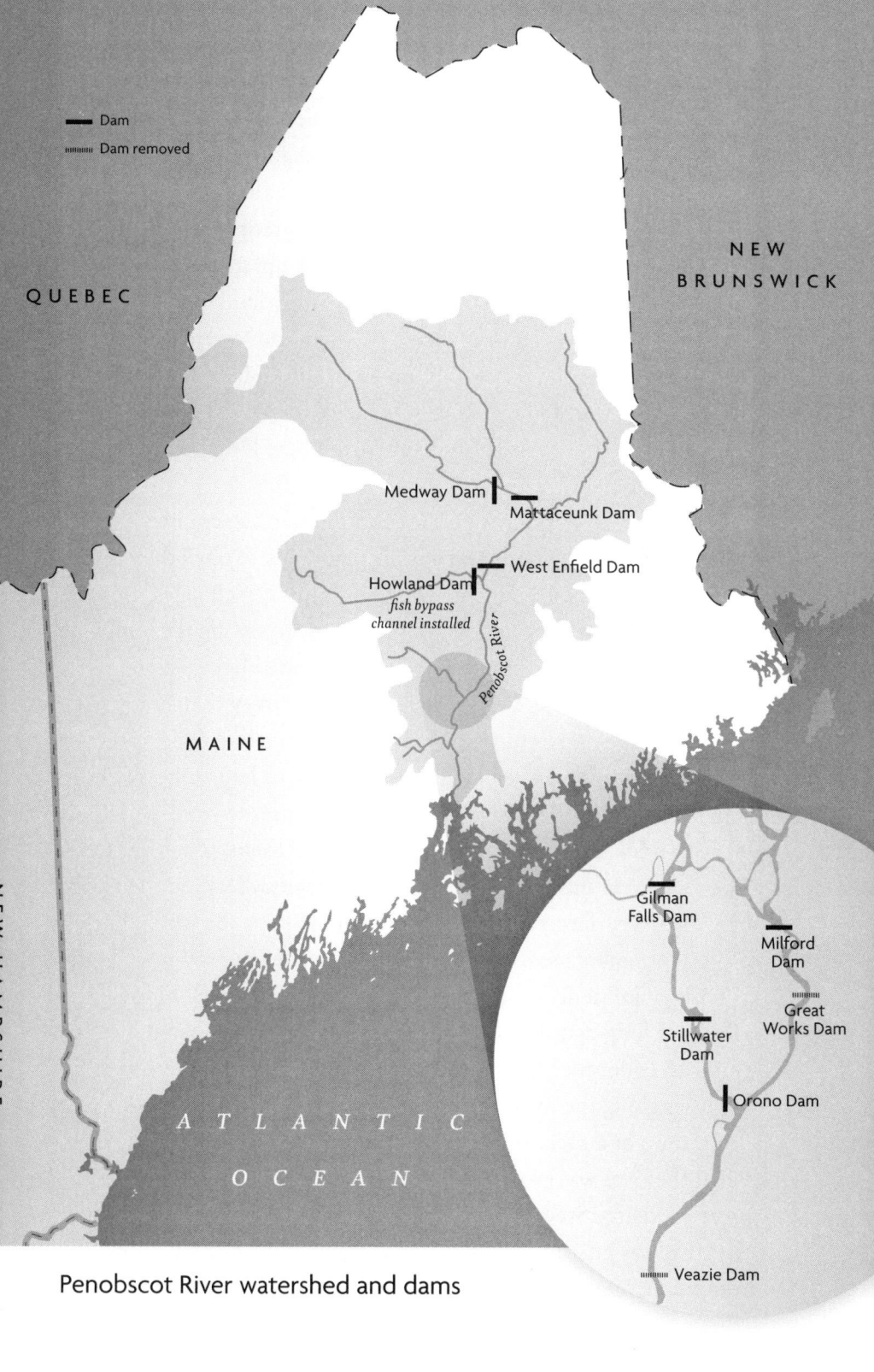

Penobscot River watershed and dams

hardwoods swelling buds, the warming sun slowly coaxing them open. The alewives, blueback herring, and shad starting to run, and the seals, ospreys, and cormorants coming right along with them to feast. Next would come the juvenile out-migrating fish downriver in their silvery wriggle, and the river would be in its most abundant time, the side channels full of alewives, the trees loud with migrating birds returned and defending their territories, and soon, chicks on the nest. By summer the fish would move to the Gulf of Maine. Then the ice would start to form, the bookend to the year. These were the great gyres of the seasons, forces of nature here long before us, and all they influence.

In addition to working with the echosounder, Stevens made visual counts of fish-eating birds and marine mammals, to track their distribution and abundance. "It's amazing how rapidly it changes," he said. Temperature, not the calendar, was the maestro setting the tempo of this orchestral composition of lives, intertwined between the river, the land, and the sea. Two weeks before our survey that sunny May morning, the temperature was four to five degrees colder, and on a previous survey that week Saunders and Stevens had witnessed little activity. But with the nudge of warmth now, ospreys were paired up on their nests and cormorants were rebuilding theirs, piling on sticks. The seals were more active. New life was everywhere: on the water, beneath the surface, roosting in trees, and prowling the banks. Eagles, osprey, cormorants, seals, and the great masses of fish, moving at depth.

History influences this river. While its scars are well hidden, the effects of long-ago damage continue. Stevens remembered a canoe trip with colleagues, and how easy the paddling was—there were no big trees along the river, and the river was simplified, with not a tree or a stump or a rock. It was made that way during the logging days, when the river was groomed for moving millions of logs. It sure was great canoeing, Stevens said. Nice and easy. "We were being taught that this was pristine," he said with a shake of his head. It shows the communication challenge, he noted, when even scientists were not aware of the impacts of historical practices. "What would these rivers be like if they had not transported logs, and they still had continuity with the floodplain? The East Coast was destroyed for three generations before we went west to mess *it* up. What we have here is artificial, our grandparents would not even remember an intact river, they were on their second and third cuts by then. It's a disappearing baseline."

Another hidden scar is on the bottom: the riverbed is smothered with a massive load of woodchips, left from the days of the sawmills, when waste was dumped right in the river. "The scars are so hidden, you have no idea what is underneath you," Stevens said. "And there hasn't been an active sawmill on the river in fifty, seventy, eighty years. You bring that stuff up from the bottom and it is like it came off a chainsaw yesterday." The bottom of the Penobscot is so problematic, they gave up sampling it using an underwater sled and instead stick with using sonar in the water column. Stevens is amazed that there are native species left here at all. Not long ago, the river was so anoxic, and so polluted, it was different colors, he said. "That has affected how a lot of people think about the river. For the most part there was not a natural draw to the river to make it what it should be, it was more a memory of a river that is polluted." Yet the Penobscot is coming back quickly.

The data they have recorded over time shows an explosion of life, roughly three times the fish as when they began the survey, Saunders said. In general, they were finding that the numbers and size of fish were both greater and less patchy after dam removal. And fish-eating birds and marine mammals were present in greater numbers too. The estuary is a tough place to live, with its changing tides, freshwater river flows, and weather—this is one of the most dynamic habitats for aquatic life anywhere. But the estuary also offers the rewards of rich food coming from both the aquatic and freshwater environments in this vast mixing zone.

Overhead a bald eagle was being harassed by a crow. A bait ball—a cluster of forage fish—in the middle of the river was setting off a feeding frenzy of seabirds. "The birds and seals know where the fish are. And they are there because that is where the zooplankton are," Stevens said. A red fox wandered the shoreline. We all felt lucky to be out here, seeing this resurgence. "It is a lot of people's hard work," Saunders said. "Landowners, select boards, local politicians. To see that ecological restoration is possible if we give these systems a chance. It's a hunch we have been working most of our lives, and to see these fish respond, it's just a really good feeling." Saunders didn't know it then, but before spring was out, nearly six million river herring would blast up the river—double the record set just last year. "River herring, they are doing it, you just give them a chance and they are back," Stevens said. "They are driving everything else, that is why we see the cormorants, the seals." It was naïve to think salmon—with a system so much more broken, over a longer migration, and still facing seven or eight dams with inefficient fish passage—could already come surging back too. More work needs to be done before that can happen, including better passage at the town of Milford.

There have been hairy moments along the way in his career as a fish biologist, Stevens said. Like driving back in the van borrowed from the university motor pool to campus at the University of Maine in the middle of the night, after a long day of fieldwork, and hearing a funny sound. He turned his head to see eels had escaped a cooler in the back and were climbing the walls of the van, swimming in their slime. "It was, 'What do we do about this van? We have to bring it back to the motor pool.' I was just eighteen, nineteen, it was a work-study job."

Saunders snapped him out of reliving the slime: "Whoa, look at that cloud," he said, pointing to a swarm of dots on the screen. "That's what we are talking about." Fish. Lots and lots of fish. Historically, fish runs on the Penobscot were spectacular, with an estimated fourteen million to twenty million alewives, a species of herring returning to the river, along with seventy-five thousand to one hundred thousand Atlantic salmon and three million to five million American shad, according to NOAA. This was a river teeming with migratory fish, not only salmon and river herring and shad but eels, sturgeon, tomcod, rainbow smelt, sea lamprey, shortnose sturgeon, and striped bass.

Yet many of these fish populations are now at all-time lows, or below historical estimates. Commercial fisheries for shad were shuttered when the species declined to the point of extirpation in many rivers in Maine. Atlantic salmon and shortnose sturgeon are listed as endangered under the Endangered Species Act, and Atlantic sturgeon were listed as threatened within the Gulf of Maine. These grim declines were the main impetus for the ongoing restoration work on the Penobscot River. In the United States, Atlantic salmon historically ranged from the Canadian border to as far south as Long Island, which extends from the New York Harbor east into the North Atlantic Ocean. Today the last of those wild populations barely hangs on in the Penobscot. Atlantic salmon remain at very high risk of extinction. The Atlantic salmon of the Gulf of Maine are, of all the animals NOAA protects, one of only ten Species in the Spotlight, meaning they are among the most endangered. Overfishing, pollution, egregious habitat destruction, and especially the more than one hundred dams built beginning as early as the 1830s all over this river basin have all imperiled the native fish of the Penobscot. Known as The Leaper for its ability to jump as high as twelve feet, even an Atlantic salmon is no match for a concrete wall.

Today the $64 million Penobscot River Restoration Project is among the largest and most ambitious river restorations underway in the United States. It got started with a historic settlement agreement to initiate the restoration plan, signed by the Penobscot Nation, five non-profits, and state and federal agencies in 2004. This was a watershed agreement in more ways than one: negotiators mastered a compromise that improved fish passage while increasing the amount of hydropower generated on the river. Dam removal began in June 2012, taking out the Great Works Dam between Old Town and Bradley, then the Veazie Dam between Veazie and Eddington in 2013. Also accomplished with deft compromise was the creation of the Howland Dam fish bypass channel on the Piscataquis River at its confluence with the Penobscot. The bypass, completed in 2016, lets the fish go around a decommissioned dam the community wanted to retain, for the sake of lakefront properties and recreation. A new fish lift also was installed at the Milford Dam at river mile 38 in 2014 to improve fish passage.

It's a contraption informally known at The Salmonator. In May 2023, when the salmon run was on, I went to see it. Essentially an elevator, this lift raises fish up to pass over the dam. As I faced the lift, the operator opened a gate and out gushed water roiling with river herring. In the thrashing mass, one fish stood out with its silver swagger. Atlantic salmon don't look like anything else. So much bigger, stronger, and all supple muscle—this was no herring. Technicians from Brookfield Renewable Partners, a Canadian corporation that owns the dam and sells its energy, saw the fish and operated the lift

This fish lift at the Milford Dam gives fish, including river herring and one Atlantic salmon (visible at center) a ride up and over the dam on the Penobscot River.

like footmen for royalty, this King of Fish. The salmon was carefully sorted from the mass with a gate and scooped into a holding tank. There, another dozen or so silvery salmon circled until they would be taken to a hatchery for spawning. Hatcheries have been crucial for conserving this species, so close to the oblivion of extinction. The fish lift is part of improved access to 368.5 miles of the Penobscot. Today there is open, unobstructed passage to 14 miles of the river, including lower tributaries.

The effort has been helpful for fish recovery, but it is just a start, said John Kocik, chief of Atlantic salmon ecosystems research for NOAA Fisheries. Hundreds of barriers including culverts and dams still block fish migration in the Penobscot watershed, and passage even where it has been provided is under scrutiny. At Milford too many endangered salmon are waiting too long to cross at the fish lift, a violation of the company's federal license that will have to be addressed. Better, faster passage is mandatory.

Juvenile fish in particular suffer in their downstream migrations, when dealing with fishways and other devices intended to get them past dams. More work is necessary for the restoration plan to live up to its potential to increase access to spawning habitat and achieve recovery of sustainable populations of migratory fish in the Penobscot. Tellingly, a decade after dam removal, a 2023 paper found that upstream of the lowermost dam the river remains dominated by lake species, while adult sea-run fish continue to be most abundant immediately downstream of the lowermost dam. Benefits for one species will improve survival of other sea-run fish in this braided river of life. The Penobscot is not like the great Pacific salmon rivers of the Pacific Northwest, historically home to runs of salmon counted in the millions. Rather the Penobscot is a *herring* river, which also is home to salmon. So for salmon to improve, it will take a robust, multispecies recovery effort.

The twelve runs of sea-run fish in the Penobscot live in an intertwined community of species, beautiful in their delicate interrelationships and carefully orchestrated run timing. For instance, sea lamprey make their nests in the river by carrying one rock at a time in their mouth from other locations, to deposit in a loose pile in riffles. In so doing, they create piles of clean gravel *just the right size* for a second use, by salmon making their redds, or nests. The spawned-out lamprey—which look like eels but are actually a jawless fish—also create a boost in marine-derived nutrients for the emerging Atlantic salmon, helping them to increase the size, growth, and body fat of the baby salmon, scientists found. These sorts of gorgeous interrelationships, refined over thousands of years of coevolution, exist with other species too. When the baby salmon, called smolts, are beginning their migration to the sea, they are heading downstream just as the far more numerous alewives and blueback herring are returning to the river. This, scientists think, may create a flood of silvery distractions for predators that lessens the toll on the baby salmon. Restoring this coevolved suite of migratory fish to levels that sustain these relationships is hoped to make the river come alive again in so many ways: cycling nutrients, providing prey, and creating spawning and rearing conditions that make the river not just survivable but suitable for Atlantic salmon and their fish kin. Maine is their last stand, and the Penobscot their stronghold. Helping these fish thrive in the river again is hoped to also boost the web of life at sea, where more fish out-migrating from the river means more food for cod, pollack, and haddock, in an unending cycle of renewal.

Habitat work is starting to pay off: incredibly, in 2023 some 5.5 *million* river herring were counted at the Milford Dam in Old Town, up from a mere 2,000 in 2011 before dam removal, according to the Maine Department of Marine Resources. In all, 1,705 Atlantic salmon were counted at the trap at Milford, up from just 1,076 in 2019. While these are very small numbers, it's better than in 2014, when only 248 Atlantic salmon returned to the Penobscot to spawn, according to NOAA. Thousands of shad are also passed each year above Milford. Still not a lot and still a fragile population. But consider before Veazie,

An Atlantic salmon cruises into view in the fish window at the Milford Dam. Dam removal on the Penobscot is helping to rebuild sea-run fish populations.

the lowest dam on the system, was taken out, only 16 shad passed above Veazie in the highest and best passage year ever recorded.

For all the changes they are witnessing, Saunders and Stevens know dam removal won't be enough. These fish also have to have places to go back home to that function more like they used to. "How are these systems broken, what do we need to fix, we have to think holistically," Saunders said. River recovery has become a movement that is about not only dam removal and salmon but a multispecies recovery and improving habitat conditions in the streams throughout the historic range of sea-run fish. "Land trusts, Maine Audubon, the Appalachian Mountain Club, a host of nonprofits all are getting involved, it isn't just the fishing clubs anymore that care," he said. The geography of the effort keeps expanding along with the people involved, doing work all the way into timber country to take out bad culverts, and remove the small dams blocking the ponds alewives seek. This is the tiny do-everything-everywhere community work that makes big steps like dam removal really start to go somewhere—because the fish can.

STEVEN KOENIG is associate director of Project SHARE, based in Eastport, Maine, an organization dedicated to collaborative projects to enhance salmon habitat. Part of what the group has to fight, Koenig said, is collective amnesia. Salmon and other sea-run fish have been gone for so long, people have forgotten about them. He met me along the east branch of the Machias River to show off restoration work they were doing for fish. I could see the problem: to look at it, the river seems fine, even scenic. "People think this is pristine because there are no houses. There is nothing on it, no industry," Koenig said. "But just because there are no houses doesn't mean it's okay." Dam removal and culvert repairs to restore connectivity of the river for fish is only part of the fish restoration task; the tributaries also have to be in shape to nurture those fish once they get there.

Just as on the main stem of the Penobscot River, here on the Machias, miles of streams had been damaged, compacted and flattened, their banks ripped straight, the better to use them as sluiceways for logs. No attempts had been made to restore the streams after these assaults, which destroyed shade trees along the banks of brooks as well as the small plants and bushes that served as bank cover, and even the banks themselves, allowing water temperatures to climb in summer. Large boulders, logs, and debris had been cleared completely from stream channels to create a smooth run for the log drives. Pools in the streambed—so necessary for fish survival for resting and feeding and rearing—had been filled in and any turns and bends and oxbows (all the complexity that defines good fish habitat) had been systematically eliminated. This had been done to brooks and streams all over the watershed. Stream bottoms had been mashed flat, and the overhanging and bank vegetation that produced food for fish also had been destroyed. Clean gravel needed by everything from eels to salmon for building their nests had been buried in mud and silt. The widening of streams by this alteration and simplification had shallowed the water, making egg freezing more likely—even as it hastened runoff in the spring, and made for lower flows in summer, worsening the environment for fish in every season.

Dams on these rivers were operated for industrial purposes—moving wood or generating power or both—without a thought as to their effect in ramping up and down water levels on trout or salmon fisheries. Fish were stranded in small pools when water levels dropped downstream as dam operators closed their gates to pond water to make power or get ready to move wood. Eggs dried on banks suddenly dewatered. Production of aquatic insects cratered, and what living fish remained were crowded in small pools. Yet the log drives were just a warm-up for the damage done by modern equipment, with extensive road building and trucking over haul roads, failed culverts, and cascades of sediment from poorly maintained roads—all lethal for fish. The heavy cutting itself in the watershed, especially along lake shores and brook banks, changed the hydrology of the forest. Clear-cutting and building haul roads and landings for stacked wood to be launched into the stream for the spring drive all made for quick runoff and siltation. The legacy of all this, Koenig said, is that brooks, streams, and rivers even in relatively undeveloped drainages—without houses, towns, buildings, or people—are deeply damaged. What looks just fine to the untrained eye is a severely compromised environment. "It will take seventy-five years for the forests to get old enough for wood to fall in and begin these natural processes," Koenig said of the East Machias. "That's a generation before this river can once again begin to function on its own, accruing big wood, building jams, complexity, and fish habitat."

Chris Frederico, executive director for Project SHARE, stood on the banks where one of the group's projects has been putting big wood back in the stream, to rebuild complexity so relentlessly removed by the log drives. "We call it chop and drop," he said, showing where trees are cut on the banks here and there, and carefully felled in the stream. It took a week to fell and place just eight trees, carefully cutting the roots with a Sawzall then using a grip hoist and a block and tackle to put the trees in the river. Other wood is hauled to the site and placed with heavy equipment, and pinned in place to keep it from washing away. It's work literally one log and root wad at a time, to restore the ability of the river to nurture sea-run fish. He knows it will take a long time, and continued work, to keep feeding this river starved for so long for big wood. "We have to be patient," Federico said. "This river is hungry."

It's working. Dozens of fixes all over the watershed are complementing the big dam removals and passage improvements to recover rivers, streams, and watersheds to undo the damage of the past. Tributaries of the Penobscot such as Sedgeunkedunk Stream in Orrington, Maine, had no alewives at all *for nearly two centuries* because their migration was blocked. The town of Orrington and multiple partners, including the Penobscot Nation, worked to fix that. "It was shopping carts, needles and tires, an urbanized junk pile," Dan McCaw, the fisheries program director for the Penobscot Nation, said of the stream. But, as he told me, the system is resilient:

> If you build it, they will come. Alewives don't care. You give them an eighth of a chance, they will do it. Historically, the populations were massive. The energy they are bringing in their carcasses, and the sea lamprey too, that's more bugs in the system, the more juveniles you can raise, the more nutrition you are putting back. Every little tributary adds strength and resilience to the system. Alewives are the biomass bomb of the river, feeding the land and the river and the sea. Everything eats alewives, at every life stage. They may not be very big, but nutritionally, they are building blocks of the ecosystem. Getting them back is key to the recovery of every other species, including Atlantics.

It's nice when you come into this in the spring and it smells like fish. And it is raucous with life. And the racoons, the black flies, everything is eating. The smell of rotting fish is perfume to people who understand what it means. It is great, it means it's alive. All the creatures are coming out to eat. I once saw a mink fighting an alewife almost as big as it, to drag it back to its hole.

I wanted to see this for myself. So in the spring, McCaw said he knew just where to take me. How beautiful that this is something people now do in Maine: schoolkids by the busload, families, people like me, just amazed to see the survival and persistence of this fish, going out on spring days to feel the awe. In May 2023 we met at the fishway at Blackman Stream at Leonard's Mills Maine Forest and Logging Museum in Bradley. And there they were: thrashing their way to Chemo Pond. "Hello, little darlings," McCaw said, the delight real in his voice. Ten- to twelve-inches long, all muscle, shining silver, their fins knifing the water, the alewives were oblivious to us, surging upstream, their animal determination exhilarating and sublime. Uncountable, unstoppable, they zoomed by in an unbroken stream.

That night, I hopped back in the car to go see where they came from, in the Lower Penobscot River, at the former site of the Veazie Dam, where these fish now had unimpeded access for their final push to their natal stream. I arrived at twilight, and the sinking sun was pinking the water. Shadbush glowed with white blossoms. The river murmured over riverbed rocks aswirl in the blue, silver, and lavender of its waters, reflecting the evening light. A dipper dunked up and down at the water's edge, and flitted across the free-flowing current. It was hard to imagine now, watching this river, the dam had ever been here at all. Across the river a chunk of concrete abutment was all that was left—that, and a monument to the dam, set up on a grassy bank. A dam turbine, set for posterity on a concrete pad.

For all that's been accomplished in the restoration of this river now so fully underway, it almost didn't happen. John Banks, the natural resources director of the Penobscot Nation for three decades, started

John Banks, former natural resources director for the Penobscot Nation, holds a war club gifted to him for his work on dam removal on the Penobscot River.

working for his tribe in humble circumstances. "There wasn't an office, not even plants," he said, as we talked at his home in Orono, Maine. Growing up in the same house as his grandmother, Cecelia Ketchum Banks, Banks was among the last of his generation to live with a fluent speaker of their Native language. People's lives were changing fast: his father was an engineer, designing nuclear weapons systems for Trident submarines. It was an uncle who encouraged Banks to get a forestry degree so he could do what he loved to do—spend time in the woods—but still get a good job. Banks graduated from the University of Maine with a degree in forestry in 1980. It was a tumultuous time for the tribe. The Penobscot had just signed the Land Claims Settlement Act, extinguishing their aboriginal claim to millions of acres of land in exchange for $82 million. Banks was just twenty-six when the tribe hired him to begin its natural resource program. "I had to build it from scratch. When I got there, there was a forest technician and a Jeep," he said.

The tribe manages its lands and waters not only for science, and for revenue for their government, but from a cultural perspective. "Those are our relatives. Management is for seven generations in mind," Banks said. "The forest has been the source of our sustenance for thousands of years. It provided the food and shelter and warmth in winter. It provided everything. Tools, weapons, bows, arrows, everything, all the needs were met." In the Wabanaki origin story, people are born from a tree, the brown ash. At the community center on the reservation on Indian Island, Banks ran his hand down the rail of a canoe, one of several put away for the winter. Handmade from birch bark, with spruce for the thwarts and rails, it would be used for the tribe's summer paddle in the Penobscot to Mount Katahdin. "We had such a close relationship with the forest," said Banks of tribal members, and they still do. That relationship extends to all the living things in it and sheltered by it. Everything that flies, swims, walks, or crawls. Including salmon. Salmon are creatures of the forest. The Penobscot are a salmon people, and they are healing the lands and waters of their ancestral territory.

In 1988, Banks and Jim Sappier, Penobscot tribal chief, in a dawn ceremony committed to restoring Atlantic salmon to these waters. "We said prayers for them to continue to come back home," Banks said of two fish caught by tribal members for the ceremony, under a permit issued by him. Banks and Sappier prayed not only for the salmon but the eleven other runs of sea-run fish they coexist with in the river. There were many times the agreement that launched the restoration stalled. Banks remembered one grim meeting as negotiations got lost in the weeds and were grinding to a standstill. "These attorneys were arguing and taking up time. Arguing about whether this word is a word or not. And I could see that they were really going at it with each other, and they were fighting over at the tribe too, and the NGOs, and I was being pulled. I had to do something, I didn't know what." Banks brought an eagle feather to the next meeting.

> I got my eagle feather out, I asked for time at the beginning of the meeting. I didn't know what I was going to do at all, but when you get into these things, something happens at a certain point, and you are not in control any more. Whatever is happening is happening through you, you are just a vehicle, and that is what happened. I went around and started talking. Touching each person on the shoulder with the feather and talking about "You think this is about you and your career; this is about the river and the ones that don't have a voice; don't lose sight of that." And I could feel a different energy coming through, and when I got to each person, the words were what that person needed to hear.

As important as the agreement was for fish, to Banks even more crucial was that breakthrough, to show that people of different walks of life around a table—from state and federal agencies, utilities, fishing groups, community members, the Penobscot Nation, and environmental groups—could find a solution for the benefit of something they all care about. "That is the lesson, that is what matters most. One aspect of this was to show that if people want to work together, they can. That was the big thing that gave me hope, that a group of dedicated people can come together to try to fix something. That is what you can learn from the Penobscot project, that if people put their differences aside, and work on what they have a common vision to do. So many people came together with a vision for the future health of the watershed." The benefits go beyond the river to the larger landscape, Banks said. "It's bearing fruit, people are starting to come around to understand there are other values with the forest other than just making money cutting trees. Recognizing there are habitat values, and carbon sequestration values. I think it is just persistence. Staying calm. I remember one of my mentors, a tribal elder telling me as soon as you lost your temper, you lost the argument."

Change is possible. The work on the Penobscot River and its tributaries is proving that. So does Bill London's work in Maine's northern forest, repairing the logging roads he built for sea-run fish people thought were gone for good. So does the work of the AMC, working with Molly and Alex London and their crews to restore the zombie forests of monoculture industrial paper lands. So does the work Charles Loring Jr. and Ben Stevens showed me on the Penobscot's trust land, in logging practices that while earning some money, also took care of the animals, the water quality, and left the best trees behind to get bigger. The Salmon Parks coming to Yuquot/Nootka Island, the Northwest Forest Plan that saved millions of acres of old-growth—even if it didn't rescue the spotted owl—all of these show we can rethink even deeply entrenched, destructive practices. The question in my mind is whether more of these sorts of changes can happen, at big enough scale, quickly enough. To save the best of what's left, fix what we can, and stop wrecking more. The salmon and the forests are speaking. And what they are telling us is we are about out of time. Today the old-growth trees, these last precious remnants of the grandeur that used to be, are at risk. Even where they are protected.

10

THE TREES ARE SPEAKING

OLD-GROWTH in the northeastern United States is vanishingly rare, and wherever truly old trees abide, they are always a remnant of pure chance and deliberate conservation. A roll of topography was just a bit too rough to deal with. Or a perhaps a family feud or crisis or transition in ownership hiccupped a cutting cycle that took everything else. Whatever the case, these remaining old stands, wherever they are, are there for a reason. Which gives them another layer of specialness: they are cherished, unique legacies set aside for preservation in landscapes so vastly and continually changed. Northeastern US old-growth forests look different from the towering, moss- and fern-draped forests of the Pacific Northwest. The species are different, in forests dominated by sugar maple, beech, eastern hemlock, birch, oak, and white pine. These species, even at great age, don't grow nearly as tall as the Pacific Northwest's Douglas fir, spruce, and western redcedars. And these are not swaths of forests, functioning at a landscape and watershed scale. Mostly what is left in the Northeast are scraps and pockets of old-growth, tucked away here and there.

"You come here and it doesn't overwhelm you," Neil Pederson, a senior ecologist at the Harvard Forest, said of East Coast old-growth. He has compiled a guide to identifying these elders. The grizzled bark of these trees is a hint, there is that oldness in their mien. Something about the crabbed reach of their broken stag-headed tops depicts centuries of endurance and adaptation to the neighboring trees they have long abided with—or long since outlived. Old trees are that way: their shape depicts the stories of their life. The ice storms, the tornados, the heavy wet snows of shoulder seasons that broke off their branches and tops. The shape they took to reach around a neighbor, seeking the sun. The forest around these old trees tells the story of the survivors too—the broken snag nearby or stump that shows the lucky twist of fate that leaves these old-growth denizens among the last trees still standing. This is part of the enchantment of old-growth anywhere. It is a living history of deep time and connection to cadences we short-lived humans can only imagine.

Landscape historian Mike Kudish remembers the first time he saw the old-growth stands in the Big Indian Wilderness Area, a 34,000-acre forest in New York's Catskill Forest Preserve. It's a stony, hilly, rough piece of ground between the sites of two nineteenth-century tanneries. And yet—there it is, a remnant patch of old-growth forest, with gnarled sugar maples and yellow birch, some more than two centuries and even three centuries old. "I thought, it can't be first growth," Kudish said, at a ceremony held in the late fall of 2022, to add stands in the Big Indian Wilderness to the Old Growth Forest Network. About 22,280 acres, or 66 percent of the forest accessioned, had never been altered by human activity—not logging or even bark-peeling, agriculture, burning, charcoal making, or mining. Kudish is a master draftsman of hand-drawn and -colored landscape maps of fine-grained attention. His rendering of this landscape in colored pencil shows just how unlikely it is that any original forest abides here. Threaded by brooks, there was a sawmill not far away, running on High Falls Brook as late as 1880. Tanneries operated from 1848 to 1879, Kudish noted on his map. Logging roads and bark roads to the tanneries cut through the landscape and speak to the logging that went on all around. Everywhere are houses and farms and roads—including the state highway that bisects this portion of the Big Indian Wilderness Area. And yet. Just a short hike from the highway, there the old trees still stand.

"It is amazing the logging and loggers didn't make it here," said Pederson, as we walked the site in the fall of 2022. "I had the same reaction, I can't believe this is old-growth." Yet he has confirmed yellow birch at 250 to 300 years old, even a sugar maple at 310 years old in these woods. The study of old-growth in the Northeast is complicated by its very rarity. After four hundred years of intense logging, less than 1 percent of the original forests of the Northeast remain uncut. Pederson and David Orwig, another senior ecologist at the Harvard Forest, are co-leads on a project looking deep into the past of these eastern old-growth forests. One of their first tasks has been just defining and finding them. "We are still trying to get a handle on what these forests should look like, we are in our infancy. We don't know how old they are, how they grow, we are a little bewildered. We just don't know that much about these forests," Pederson said. And old trees have a way of fooling even the most experienced eye. The big trees are very often not the oldest. The ones that struggled to get into the canopy, the ones with all the damage, burls, and twists—those are likely candidates. Top-heaviness, with most of the foliage loaded in the canopy like a bunch of celery, is another clue, Pederson said. Some species take on different characteristics in their bark as they come into great age. Red oak gets a powdered-sugar look to its bark, he noted.

There is another interesting voice speaking up in these forests. "The lichens are telling the truth. They just don't exist where you don't have the old trees," said amateur lichenologist John Franklin. He had joined a walk through the Big Indian Wilderness to celebrate this newest accession by the Old-Growth Forest Network, which works to identify and protect old-growth forests across the country. Dozens of tree aficionados including myself—from academia, state and local agencies, conservation groups, and outdoor enthusiasts—were along for a hike to explore and enjoy these woods. It was a luminous fall day as only New England can deliver, with sun streaming through the open deciduous canopy that had already lost much of its leaves. The trail was slippery with newly fallen leaves deep enough to kick through, releasing a marvelous woodsy fragrance as we crunched along. In time, the trail dropped to the crossing of Upper Biscuit Brook. Here were many larger trees, nearly three feet or more in diameter. Franklin, the lichen fan, excitedly got out a hand lens to zoom in on the bark of a sugar maple. A lichen had caught his eye. To most people, it would just be a green splat on a tree trunk. But to him, this was a rare specimen worth close inspection. Because of its size, big as a Frisbee, it was probably old, Franklin guessed, abiding with this tree for a good long time, maybe a century. The garden of lichens, mosses, and leafy liverworts on the trees were all good indicators of old-growth trees, Franklin said.

It was rather marvelous, watching these dozens of people circling the big busted-up trees by the brook, with their broken tops, cavities in their branches, and trunks sometimes caved in at the base—yet still very much alive. They were peering closely at lichens and awed by their coexistence with these trees through the centuries. How very fine, to encounter not only these old trees but also people so excited to be with them. All around, they were spending quiet moments just savoring life from the lichen and liverwort's perspective. Isn't that also what old-growth does, for people? It is a gift of awe, of wonder, just to be in an old forest. Apart from all of their ecological gifts— the carbon locked away in these trees, the water they conserve and clean, the clean air they breathe out for us, the animals they shelter—these old-growth forests also *are a refuge for people.* Here was an island of peace and perspective, a place to ponder the cycles of time, the brook freezing and thawing through hundreds of winters, beaver dams coming and going, and species carrying on their evolutionary mission. It was sobering to realize the losses over the lifetime of these trees: passenger pigeons used to be loud in their branches—but now are silenced by extinction.

In this forest the classic old-growth is in lower spots such as this, below three thousand feet in elevation, sheltered from the most severe weather. Here, the soil depth, drainage, moisture, and nutrients are optimal for a tree to live a long life. But there's no telling with northeastern old-growth. I've hiked the severe, rocky redoubt of Wachusett Mountain State Reservation in Worcester County, Massachusetts, renowned for its remaining patches of old-growth trees. Here was a mountaintop blasted by wind, snow, rain, and ice storms as well as hurricanes, its trails a rocky scramble. I wondered, as I hiked, "Where are all the old-growth red oaks, yellow birch, and hemlock I'd heard so much about?" All around turned out to be the answer. But so crabbed and tiny, stunted, hunched, and dwarfed were the trees because of their growing conditions, I never spotted them. The very trail I hiked at Wachusett Mountain, the Old Indian Trail, is presided over by a red oak determined to be 322 years old, and two others, both 320 years old, in 1996. These trees are still there today, Dave Orwig confirmed, at 350 and 348 years old—very much still the oldest known in the world for their species. Here was a forest home to venerable yellow birch, black birch, and the oldest known northern red oak *in the world*. And I had walked right by them!

It was this discovery by Orwig, at the Harvard Forest, and his collaborators of this truly unique treasure of trees and advocacy by other big tree hunters and environmental groups that in 1995 killed a plan for expansion of the nearby ski area that would have cut through these trees. Today these trees are protected in a two-hundred-acre state reserve of extremely rare old-growth forest, the largest known area in Massachusetts of trees that have never been cut. They are a cherished rare woodland attraction, featured in hikes and outdoor education programs on Wachusett Mountain. Yet even though the scientists and advocates for these trees had saved this forest from sacrifice to a ski run, this protected old-growth is still at risk. The overstory beech trees—the big ones that had claimed and held their place in the canopy for so long—were heavily infested with beech bark disease, caused by the combination of the scale insect *Cryptococcus fagisuga* and the pathogenic fungi *Neonectria*, Orwig and his collaborators warned in a report on these woods.

So often in these Northeastern woods, this is how it goes. To the untrained eye, everything looks fine, beautiful even. But to those who know, the presence of absence is everywhere. There were no big beech anymore in the forest at the Big Indian Wilderness. A third of that forest, *even in protected wilderness*, is in decline because of beech bark disease. A warty pox, the disease has left forests like these bereft of an entire generation of elders—and a terrible scarring on the bark of the younger trees yet to succumb. But succumb, they will. Death and disease stalk these woods. All over the Northeast, even forests like these that have never been cut and are forever protected from development are at risk, because of exotic pests and diseases for which people are responsible.

This is an environmental crisis people are not used to thinking about; there are no chainsaws, no bulldozers, no flames. But the damage being done by the invasive pests entering the nation's forests over the past 150 years is, along with many other problems associated with climate change, among the most profound disturbances forests face. Millions upon millions of trees, and important forest characteristics, from habitat suitability for native species, to hydrology and the interconnected biodiversity of forests, are affected by this silent scourge of invasive bugs and disease. Billions of chewing mouths are tunneling into bark, devouring the life-giving cambium beneath it that carries the vital fluids of trees and biting into needles and leaves to steal the stores of sugars trees have so diligently made, the source of all their new growth, essential to their survival. Wind and rain have been made into vectors of death, blowing spores and splashing pestilence.

It's not that trees are weaklings. Trees are miraculous, resilient beings, masters of adaptation and survival. They are autotrophs, able to make their own food themselves out of thin air by photosynthesis, a process so complex, no scientist has yet been able to replicate it in a lab. And tough? Trees are built to take a lifetime of abuse. Their leaves withstand wind and searing sun, and trees can bend not break even in roaring wind. In winter, trees can slough off glittering, deadly casings of ice on their branches and dump snow from their boughs. Only the heaviest wet snows and ice storms will rip them to pieces. But even then, surviving trees will grow new branches and surge into gaps created by their fallen neighbors. Legions of young trees will spring up and grow in the sunny gaps that beckon new life.

I've seen many old trees that are mostly dead, died-back, broken, lightning struck, burned, their heartwood caved and rotted. Yet such trees—so valuable for wildlife—persist, a shred of cambium under the bark still carrying life to the branches green and thriving even in their shattered crown. The bark on some species is thick enough to endure fire, and many can quickly resprout from cut stumps. Even in punishing droughts, trees can die-back and resize to survive, then surge with new growth in better times. When it's really bad, in understories too dark to make food, trees can just wait, zombie-like for decades, not even growing one ring, until it's their turn in the sun, and a gap in the canopy releases them to a surge of growth. Trees even practice a form of predator control, putting out variably-sized seed crops in order to defeat the many animals that will eat them. Oaks, maples, and other species deploy these feast-or-famine seed strategies, throwing out a bonanza of acorns and helicopters only occasionally, to swamp the ability of greedy mouths to eat them all. These fiestas in the forest boost wildlife size and births. And putting out such a glut—more than the animal population sized to measlier rations can eat—means there will be plenty of leftovers to seed the next generation.

Trees are consummate diplomats too. As much competition as there is between trees in the forest, as scientist Suzanne Simard has shown working in Vancouver Island on the Mother Tree Project, trees also share nutrients and water through their fungal network belowground—and even preferentially provision their own offspring. Trees even have their own version of a community defense system against the familiar insects with which they have coevolved, emitting chemicals from their leaves called pheromones, to alert their neighbors to make their leaves unpalatable. So where our native trees are dying, don't blame them. They are geniuses of adaptation to the millennia of conditions they coevolved with. But that is exactly the problem.

David Orwig, senior ecologist at the Harvard Forest, works on a survey of old-growth forest in the Adirondacks. Even after decades of study, he is enchanted by the special qualities of very old trees.

Today's catastrophic disruptions are from *novel* insects and diseases, alien invaders, against which native trees have no defense—and from which they cannot run away. This is the storm that never passes, the drought that does not relent, the insect they cannot rebuff. These assaults of pests and disease know no boundaries, and increasingly, with warming winters, no killing season. Climate warming is hastening their spread across the landscape.

I remember well meeting Orwig one winter morning, as he came back from an investigation of eastern hemlocks in the woods at the Harvard Forest that he has tracked for decades. He had been out on a morning walk, to check on the effect of a recent cold snap. He took out a sprig of hemlock he had brought back in his pocket, flipped it over, and shook his head at the thick flocking of white fluff he showed me on the underside of the twig. These tufts of white were woolly adelgid (pronounced a-del-jid), an aphid-like invader that sucks the life out of eastern hemlock. The adelgid were doing just fine, despite what we had hoped would be killing cold. Not even close. The cold was cold all right, but it would take more than just a few days to knock back the infestation hammering these hemlocks. Exotic pests and diseases are not a new problem, as the sad history of the loss of the beloved American chestnut and American elm attest. Chestnut blight originated in East and Southeast Asia and was introduced into Europe and North America in the early 1900s. That disease wiped out a huge, majestic tree with straight trunks of wood so sound and durable and workable that it was the material of choice for everything made of wood, from cradles to coffins. It grew fast, resprouted from its roots if cut, and put on a feast of nuts savored by wildlife and people alike. The writing of naturalist Henry David Thoreau on his walks in the New England woods resounds with the happy racket of people beating the trunks of massive chestnut trees, to bring down fat, succulent nuts.

But this perfect tree today exists only as a shrub, invariably dying back because of blight as soon as it resprouts and begins to grow again. Spread by spores carried on the wind and splashed by rain, the disease even turns native insects and birds into vectors of this foreign disease. The details are ghastly: A wound parasite, the disease prospers in the slightest nick exposing the live inner bark of the tree. Fungal spores develop mycelia that rapidly penetrate the inner bark and cambial layer, and eventually burst through the outer bark with pimple-like pustules that erupt through the surface of the bark. In damp weather, long twisted spore horns ooze out of pustules on infected trees, where the disease is spread by birds, crawling and flying insects, and even splashing rain. Other spores actually launch themselves from the pustules, where they are borne on the wind and can be carried for miles. This is a vision too horrible to invent, a nightmare disease that felled a dream tree throughout its entire native range. First observed killing chestnuts in 1903 in the New York Zoological Park, the blight may have been here earlier. Spreading at the rate of fifty miles per year, it was found to be introduced on nursery stock imported from Asia, eventually killing the equivalent of some *nine million acres* of American chestnut forest. Work continues to create a blight-resistant chestnut.

The near total demise of the vaselike American elm, with its graceful tresses of leaves, was just as devastating to communities across the country that made this shade tree the signature of their streets, town commons, and homes. Elm Street is one of the more common addresses in towns all over the United States for a reason. The invasive fungal pathogen put an end to that. Spread by bark beetles, Dutch elm disease—actually of Asiatic origin—was introduced in the country in the early 1930s. The epidemic began when the disease was carried to the United States from Europe (where it had already been introduced from Asia) on elm logs imported by furniture makers to craft attractive burled veneer for tables and cabinets, wrote David Karnosky in a 1979 article published in *Environmental Conservation*. The bark beetle that spreads the disease arrived on wood at major ports and railroads, transporting the logs inland, first to Ohio, then New Jersey and New York. The result was environmental havoc coast to coast, killing more than one hundred million elms, and eradicating this stately tree from forests and cities. Individual American elms, usually maintained with costly treatments, still stand on town greens and campus lawns here and there.

You might think that after those tragic losses, both disasters inflicted on beloved native species by imported products, that the United States would have embarked on a vigorous program of inspections and quarantine to prevent such misfortune in the future. You would be wrong. What has happened instead is a catastrophic *escalation* of the introduction of non-native pathogens and insects to native forests, because of the combined effects of global trade vectoring more pests and pestilence, and climate warming that stokes their spread. Trade in live plants is recognized worldwide as a dangerous pathway for non-native plant pests, with nearly 70 percent of damaging forest insects and pathogens established between 1860 and 2006 most likely brought in on imported live plants. The quantity of plants imported tops three billion plants every year, researchers found, and that was in 2007. Yet in the one year they checked, fiscal year 2009, the same researchers found about 72 percent of infested plant shipments sailed undetected through US ports. The protection of our native species is still inadequate, despite multiple reports and peer-reviewed scientific articles pointing out the danger to native forests and the necessity for better inspections and controls. It's not just imported live plants that pose risk; wood packing material, such as crates and shipping pallets, harbor invasive insects. Moving firewood spreads disease and pests. Even rainwater from infested nurseries can carry lethal disease.

This knowledge haunts scientists working in old-growth forests in the Northeast. Already so rare, even where it is protected, old-growth is not safe from humans, with our freewheeling commerce and cooked climate. Non-native forest pests are uniquely devastating: they are the terminator of trees, the only thing known to take out entire species from our forests within just decades, with enormous consequences for everything from wildlife habitat to the hydrology of the landscape and the beauty of our cities. Beloved native species celebrated in art and literature for generations have defined landscapes as big as regions, as modest as neighborhoods, and as intimate as parks and yards. And yet we are losing them to invasive pests and disease.

In his career at the Harvard Forest, Orwig has centered much of his research on the effects of the hemlock woolly adelgid, an invasive insect that wasn't even present in the Harvard Forest when he arrived as a senior staff scientist in the 1990s. He once used the groves of hemlock at the Harvard Forest as his reference point for what a healthy hemlock forest should look like. But then the insect continued its northward march. Orwig wonders, as disease and insects take their toll, whether curation and preservation of some populations of ash, of beech, of hemlock are necessary at least at some scale. "Will people look back and say, 'I wish they had done something,'" he said. Society has got to spend what it takes to be more careful in a globalized economy. These invaders are everywhere, not only in Northeast forests. In the West, white pine blister rust is killing whitebark pine, now at risk of extinction. The Northeast has been the region more affected only because of the earlier colonization, population, and industrialization of these lands, with all of their international ports of entry. In the Northeast, where diseases and insects got their jumpstart, the ten-thousand-year run of eastern hemlock is coming to an end, and other species are on their way out or already gone. But these risks are just as real for native tree species everywhere. It is dizzying how fast they can take out even old trees in protected forests.

Hemlock Hollow is a beloved stand of hemlocks at the Harvard Forest I came to know well in my time there, entranced by this dark, enveloping grove of hemlocks, including old-growth trees that have abided for centuries. Over ten years of many visits to the Harvard Forest beginning in 2013, I watched Hemlock Hollow gradually fall apart. Eventually, forest administrators, who pride themselves on providing public access to the more than four-thousand-acre forest every day of the year, took the sad step of closing the trail through Hemlock Hollow to all but researchers because of the cascade of falling branches and dying trees. Aaron Ellison, a former senior ecologist at the Harvard Forest, and sculptor David Buckley Borden in 2016–17 created an art and science collaborative they named Hemlock Hospice, culminating in a yearlong outdoor exhibit at the Harvard Forest, including Hemlock Hollow, called the Hemlock Hospice Trail, with works of art all along it. The goal of the exhibit was to get the public thinking and feeling about the slow-moving hurricane of adelgid leveling these trees and the issues of climate change and human impact on the future of New England forests.

Orwig and other researchers are exploring the death of these trees, and all the cascades of interactions these extinctions will unleash. From water usage to the amount of light reaching the forest floor and the animals that used to shelter and feed here, all that is being altered as this forest of conifers gives way to a rising tide of young black birch. While this landscape will still be a forest, everything about it will change, as the infestation advances and the hemlocks keep dying by degrees. Henry Wadsworth Longfellow famously wrote in his epic poem *Evangeline: A Tale of Acadie*:

> This is the forest primeval. The murmuring of pines
>   and the hemlocks,
> Bearded with moss, and in garments green, indistinct
>   in the twilight,
> Stand like Druids of eld, with voices sad and prophetic.

For me these beautiful words have become an elegy for the ghosts of the glory that was, in hemlocks dead and dying where they stand. These were noble hemlocks. In these woods the species itself was resurgent, coming back from an earlier die-off thousands of years ago. Next these particular trees survived the logging and tanning operations of the settler-colonists. Now only to be taken out by yet another invader, a bug.

Today, all around the gray and shattered trunks and branches is the weirdly bright, open canopy where I had always enjoyed the hemlocks' deep shade, such a plush, velvety gloom. Instead, here thrives a cheery young stand of black birch, revving up in the sunshine now bathing the understory. Like lively young visitors in an intensive care ward, there feels to me something almost indecent in the robust, burgeoning health of the surging black birch, amid so much decline and death. Change of course is normal in forests, and extinction is too. But these are not natural deaths—neither in the cause nor in the speed of destruction, with entire species being taken out in just decades. Scientists cannot believe the losses they are witnessing over their careers: losses of trees that should persist far beyond a human lifetime. Even in protected old-growth forests, elders that have stood for centuries have already passed—way before their time. Don't think for a moment this doesn't affect the scientists who have spent their careers amid these trees.

I SPENT A FEW DAYS in the Adirondacks in northern New York with Orwig and Pederson and their team researching eastern old-growth in the fall of 2023. It was eye-opening in a disturbing way. "Most people would walk through and think, 'It's a beautiful forest.' They don't see what is on the way out," Pederson said. Sunlight filtered through a canopy of hemlock, yellow birch, sugar maple, and oak in Adirondack Park that had probably never been cut. There weren't any visible signs of people here. No buildings, no wires, no roads—not even trails. Some of the biggest, oldest trees were too wide at their trunk for human arms to reach around. Cathedral light filtered through the tops of the trees, and sun flecks glowed on deep forest duff. The quiet was velvety, dimensional. These trees, some standing for centuries, unspooled their long stories uninterrupted by the noise of people. But where were the big beech? Already gone.

"I call it a quiet pandemic," Pederson said. "We are working in forests that are in late stages of decline." In forests all over the northeastern United States the big beeches are a thing of memory only. The eastern hemlocks are gray and dying. Ash trees are succumbing to relentless attack. These losses are important, ecologically. Old trees matter, Orwig explained. "They feel different and look different, and they are different—they are rare. They have such a presence. There is that wonderment. I call them the magical forest," he said. Old-growth trees are an irreplaceable living environmental archive, older than any instrument record, Orwig noted. And big old trees are needed to help save the planet. Orwig coauthored a 2018 article that found that, across forty-eight forest plots around the world, the largest 1 percent of trees comprised on average 50 percent of the aboveground biomass. That points to just how important big old trees are as carbon sinks blunting climate change. "They are the ones storing the vast major-

The white tufts on the underside of this hemlock twig
are part of the deadly trifecta of invasive species, climate
warming, and global commerce that is killing native tree
species. The white tufts are hemlock woolly adelgid,
an invasive insect that is infesting eastern hemlock.

ity of the carbon, and they also have amazing microhabitats in their
crown, completely different plants up there you would never think
of," he said. "Flying squirrels, raptors, they love being in big old trees."
The research team was astonished by a black snake climbing an old
tree to get into its crown. Who knew snakes could climb?

Orwig and Pederson made a natural team. With a multiyear grant
from the National Science Foundation, they sought the long stories of old trees, to reveal in tree rings the story of how old trees
responded to big climatic events in the past—killing droughts and
frosts—to better understand how forests might respond to climate
extremes in the future. To do it, they were coring old-growth trees all
over the Northeast and seeking the trees' stories in their heartwood.
They are tree whisperers. And it takes careful listening to reveal a
tree's story.

Protecting mature and old forests is an essential strategy in the fight against climate change, but it has not received the attention it deserves, argues Beverly Law, professor emeritus in the Department of Forest Ecosystems and Society at Oregon State University. The world's forests pull about one-third of all human-caused carbon dioxide emissions from the atmosphere *each year*. Old-growth trees in particular store enormous amounts of carbon because of their tremendous mass above- and belowground. Yet we are losing trees by the millions to invasive insects and pathogens. In a 2020 article, lead author Kristina J. Anderson-Teixeira of the Smithsonian Conservation Biology Institute and her coauthors found that in the Blue Ridge Mountains of Virginia, one in four tree deaths in the past three decades was linked to invasive species. The Harvard Forest old-growth team feels urgency in their fieldwork. "It's the rapidity of change people don't realize. Old-growth trees that have been there for centuries are being wiped out—that is what is so sobering," Pederson said. "We are catching the last whispers of beech. You see a hemlock and think, 'Oh, I'll get that the next time,' but it may be gone."

Laura Smith, at the time a research assistant at the Harvard Forest, was interviewing an elder who happened to be a tree. In a world of screen swipes and digital bytes, her tools for inquiry were low-tech and hands-on: wood glue, sandpaper, masking tape. At the Tree Ring Lab at the Harvard Forest, Smith's subject was pp317, a core she had carefully slid from the heart of an old-growth hemlock in New York's Adirondack Park. I had watched her, weeks before, as she leaned into the tree's trunk with a borer, shouldering its handle to press it deeper. The hemlock talked back with a loud creak at each turn. That was the coring crew's second day out during a week in the Adirondacks. This was an emphatically protected forest. In 1894 the people of the state of New York declared 2.6 million acres of forestland off-limits to future cutting or any commercial use, in one of the most adamant protective decrees ever anywhere in the world, setting aside a forest reserve to remain forever wild in the state constitution. In their rarity the old-growth denizens of this forest commanded the respect—and the scientific interest—due revered survivors in a mostly cutover land.

With each turn of the borer, Smith quested through the old-growth hemlock's present to its deep-time past. Finally at the tree's heart, she paused and gently slid the tray out of the borer. Cradled in it was the core: a slender biscuit-colored wand of time and a panoramic window into the past. I leaned in close and saw the core was still moist with life. Smith paused and held the core in her palms, respecting the moment this tree's story was first revealed. She slid the core carefully into a plastic tube no bigger than a drinking straw, carried in a case with dozens of other core samples gathered by the crew that day for the long hike back to the car and the drive back to the lab. It had been a good day for the crew at the end of a summer spent working on an ambitious mission to analyze cores from some seventy-six hundred individual old-growth trees, over more than 154,000 square miles of the northeastern United States, during the next several years. This is a wider geography and deeper timescale than has been previously attempted. Scientists hope to understand what the histories of these trees may tell us about how forests will respond to future shocks. Though such predictions are just that—no tree alive today has experienced what could be coming with the combined effects of globalization, climate warming, and invasive species. It is the dance of trees together over miles and centuries that these scientists are seeking, witnessed in tree cores that record their rhythms of growth and decline.

Smith held up the core she pulled that day in New York's Adirondack Park from the hemlock pp317. Now mounted on a slender wooden block back at the dendrology, or tree core, lab at the Harvard Forest, the core was ready to work up. First with a belt sander then by hand with ever-finer sandpaper, Smith smoothed the core down, gradually revealing clearly segmented rings of annual growth. In any tree, tight and narrow rings show a tree biding its time, not doing much—something is holding it back. Wide rings show a release from constraint and growth spurts when some bonanza of fresh opportunity arises. Had a tree fallen nearby, bathing the tree in sun? Or was there an event of bigger impact and scale? One core shows these events in a particular tree's history. To get the bigger picture, the team cores multiple trees in the same study plot, then across a wider area, to look for larger patterns: the forest for the trees.

Dendrochronology—the science of studying the growth of a tree and forest dynamics through the study of tree rings—is a special science. Done correctly, the error bar in counting a tree ring is zero. Tree coring allows scientists to interrogate nonverbal subjects about the course of their lives and events in their communities. Year by year, every tree writes its autobiography, and the elders in any grove record wisdom earned over many centuries. How they got along and managed with their neighbors, the changes in their community, the days of harmony, strife, and struggle, the seasons of feast and famine. Intimate diaries of their long years are recorded, in unstinting detail. Feats of solo virtuosity and brilliant synchronous survival across vast geographies are all forever inscribed. These are epic memoirs, told in the quietest of voices: the cores of old-growth trees. They recount the agency of trees, and how they make their place. Stand their ground. Endure.

Smith slid the sanded core of pp317 under the microscope and clicked on its light. She fed the core with her right hand under its lens, and with her left, she laid down a tiny dot with a pencil on the mounting block as she scanned through tree time. Every decade got a tiny, penciled dot. Fifty years got two dots, one hundred years three. Her lips moved as Smith quietly counted off the years. "There's a century, two," Smith said. "We are back to 1850, 1800, 1774 is the earliest ring." This hemlock we had sampled has abided since before there was a United States, before there was a state of New York or an Adirondack Park. Since the First Peoples utilized this place—and since the time of their resistance against settler-colonizers who would cut most of these trees. The colonizers who are why today there are so few old-growth stands left anywhere but especially in the north-eastern region of North America. Smith said she felt the weight of interpreting the tree's story, as she recorded images of the core in the digital database the team was building. The core itself would be stored in the permanent collection archived at the Harvard Forest.

As they work the team feels the press of time to get to their rare and fragile subjects, enduring in an increasingly hostile world. After a day of hard searching last summer, the team finally almost by accident located a stand of eastern hemlock on a preserve in northwestern Pennsylvania. When the team stumbled upon the trees, senior ecologist Neil Pederson said he wept with relief. They were still there. Still standing. Still alive, for five hundred and more years, some of the oldest eastern hemlocks known. Which, given what they are up against, is remarkable. Hemlock. Ash. Beech. Chestnut. Elm. A roll call of the lost. "They are a monument to our capitalist system," Pederson said of the epidemics of invasives. "Money comes first. These

Dendrochronology allows scientists to see deep into the lives of trees. Tree cores reveal patterns of growth that tell the history of a tree and a grove.

forest systems took millennia to develop. We don't even know what we are losing."

BUT THE PEOPLE OF THE DAWNLAND DO. Wabanaki basket makers are in a race against time to pass their basketmaking tradition—one of the oldest known art forms in New England—on to the next generation, as brown ash trees, also called black ash, are wiped out in state after state by the emerald ash borer. Initially detected in southeastern Michigan and southern Ontario in 2002, the emerald ash borer, native to Asia, was probably introduced sometime in the 1990s. Today the borer is on a relentless death march across the country, killing millions of ash trees of every variety in at least thirty-six states so far. Wabanaki teachings of reciprocity and responsibility in living with all their nonhuman relations for the benefit of future generations once again confront an onslaught of invaders. For the Wabanaki the loss of the ash tree is a deep blow to their culture. The tree is uniquely suited to their basketmaking because of where and how it grows and how it can be worked. The Wabanaki call brown ash the Basket Tree. It also is at the heart of one of their creation stories.

Scientists and cultural leaders are working together to protect what ash they can across the Wabanaki homelands and prevent the spread of the ash borer. Carrying on the basketmaking tradition going forward will depend on what happens to the 2 to 3 percent of the forests that is suitable for growing basket-quality ash. The tree grows in small wetland pockets, and it has no substitute. Nothing splits or works like black ash, prized for its strength and flexibility. Its cultural and economic importance to basketmakers is immense. One of three hundred to four hundred documented invasive species already in North America, the emerald ash borer is one of the gravest threats to tribes in terms of forest pests, potentially affecting fifty-four federally recognized tribal nations.

Butch Phillips is a Penobscot elder and artist who works in birch bark. I sat with him at the annual Wabanaki Winter Market in December 2022, and talked with him as basketmakers all around him sold their wares and Phillips did a brisk trade in his moose calls, made from the bark of white birch. He etches each of these art pieces by hand, cutting through a paler outer layer of bark to the rich red-brown of the bark below. The decline of the ash tree because of an unchecked invasive species saddens him. "It's terrible, it's another example of introducing a non-native species and it's our basketmakers' livelihood," Phillips said. "We don't use any other tree for our baskets. Ash is part of our creation story. And yet here it is. On the brink of extinction."

11
HOME KEEPING

IT HAD BEEN A LONG JOURNEY. A continental ramble, from the Pacific salmon forests of the Northwest to the Atlantic salmon forests of the Northeast. Places greatly afflicted, traumatized, and contracted from their original grandeur—but still here, still alive, still full of wonder. And also, full of brave, committed, smart people. From tribal leaders saving the last of the old-growth forests in British Columbia with the Salmon Parks to a logger's dad helping sea-run fish return to Maine's inland forest streams for the first time in more than a century. What I saw everywhere I went were places of love and beauty and people who care deeply about them.

This had all started at the H. J. Andrews Experimental Forest in Oregon's Cascade Range, in that extraordinary old-growth forest. So after fifteen months of work on both sides of the US–Canada border and the North American continent, the Andrews is where I wanted to return. Just to see my tree again, the one I had climbed with Mark Schulze, director at the Andrews. To smell the forest, hear the greeting of Lookout Creek, and most of all, to walk amid those centuries-old trees on the Lookout Creek Old-Growth Trail. So in July 2023, I packed up the car one more time, and drove from my home in Seattle to the Andrews. When I arrived, the air smelled of smoke from a forest fire burning some thirty miles away. This had people at the Andrews on edge. The last miles of the drive to the Andrews are through the burned and black spires of forest roasted in the 2020 Holiday Farm Fire. One of the largest in Oregon's history, it burned 173,393 acres of forest. Every year more of the cherished old-growth in this region was burning, and each year the fires came closer.

I went over to the headquarters building in the morning, to say hello and let people know I was heading out to hike the Lookout Creek Old-Growth Trail. Laurie Good, site manager for the Andrews, handed me a KN95 mask, just in case the smoke got worse. She walked me back to the rack of two-way radios, and asked me to take one and keep it turned on. "We may need to reach you," she said. I took the radio, and promised I'd text her when I got back to headquarters. I reflected, as I drove to the trailhead, on what it felt like to be going to my place of refuge, where I find stillness, with a mask for wildfire smoke and an emergency radio. I turned it to the lowest volume and put it in my pack, where it murmured. I noticed the sky had an odd pall to it—wildfire smoke from the fire over the ridge, started by careless campers. The smoke tinted the sun amber. But I was here to enjoy the trees. So I started off on the roughly six-mile hike—and stepped into the old-growth.

I've seen a lot of old-growth forests. But the majesty on this old-growth trail is a world apart. Here was a forest on its own timescale, a sovereign kingdom of green, reigning for centuries. I let myself just be here. I did this every time I visited the Andrews, walked this trail to think about all these trees had seen. So much history, so many animals, for some of these trees, more than a thousand seasons. They were a touchstone for me, those big old trees so long abiding, in a too-fast changing world. I made photo after photo of this perfect place, so exemplary of all I had learned about forests. The gaps that I had learned were so important, the new forest rising there just as it must. The berries and ferns that fed and sheltered so many lives, the buttressed trunks of cedars with their wizened, shattered, silvery tops. Lookout Creek glinted through trees, green giants of dreams: cedars, Douglas firs, and spruce, with tops too high in the canopy to see. In the ever-shifting light, the forest floor I walked was soft, and the trail stepped me through a living herbarium of Pacific Northwest native plants. Here was something so rare in this world: a vision of completeness. Nothing missing. Nothing wrecked.

I reached a log bridge over Lookout Creek and thought about my options. This ad hoc bridge—not the classy official one with the railings at the other end of the trail—was a wilderness classic of Pacific Northwest hikes, a single log, fallen or placed across a stream that invites an adventurous passage. If I fell, I thought, no one would likely know it for quite a while; I had seen not another person all day. The radio purring in my pack would surely drown. But I wanted over, wanted to cross. So, sliding down to my rump, I crabbed across, counting each scooch to stay focused. Somewhere around twenty-seven, I had made it, and managed to right myself, turtling onto all fours, then standing up. I realized once I crossed, I wouldn't have time to finish the trail anyway to get back when I promised—and I didn't want a community already on edge to worry about me. I crabbed back—it was harder in this direction, the sloping log was uphill—and hiked out at three times the pace of my meditative walk in. I was back just in time to make my window to check in when I had promised. Snugging the radio back in its slot at the headquarters building, I thought about what a new kind of hike in the old-growth I had just had, with the threat of fire for company. I drove back home to Seattle, content in the circle I had completed from the Andrews, across thousands of miles of journeying, and back again.

Less than two weeks later, I got the news. On August 5 a single lightning strike had started a wildfire at the Andrews. I must have been one of the last people to see the Lookout Creek Old-Growth Trail still intact. The Andrews would be a firefighting war zone for months, with hundreds of firefighters cutting, bulldozing, trenching, and dousing the fire with chemicals. Fire lookout Robert Mutch had first called in the smoke. This forest had seen fire before—but he knew this time it was going to be different. The fire was in steep, difficult terrain. So much so that unlike prior starts firefighters had snuffed right away, this one they decided they could not initially safely combat. "I was just kind of watching it with dread," Mutch said. "By the second day, I knew it was not going to stop." By the time it was out two months

The Lookout Fire at the H. J. Andrews Experimental Forest started August 5, 2023, and was the first significant fire at the research station. It eventually burned more than two-thirds of the forest, including old-growth that had not seen fire in five hundred years. Photo by Robert Mutch Ecosystem Photography.

later, the Lookout Fire had burned 25,754 acres—about 68 percent of the Andrews.

It's not that the scientists didn't predict someday this would happen. Fire had been coming closer every year—but had not yet seriously burned the Andrews. But while summer drought is something these trees evolved with, the droughts were becoming hotter and longer. That July, the Andrews was in the second month of a prolonged dry season. The soils had dried to August levels, and the fuels—grasses, shrubs, dead leaves, conifer needles, and sticks scattering the forest floor—were baked to tinder. The Andrews was at the end of two weeks of intense atmospheric stress, with dry conditions sucking the water out of these trees. That set up the conditions for high-intensity fire. The era in which we used to think about old-growth forests as the asbestos forest is over.

After I heard about the fire, I made another trip to the Andrews and found the community in grief. Philosopher Kathleen Moore had been fishing in Alaska when she got the news. She is a founder of one of the celebrated humanities programs at the Andrews, centered on interpreting the forest in writing and poetry. When I visited at her house in Corvallis, the fire was still burning at the Andrews. The fillets of coho salmon she had caught and brought back from Alaska were laid out all over her kitchen, drying. We went out to eat breakfast together in her back garden. "I'm trying to think about how we're going to find this place between hope and horror, how are we going to find a way to live between these extremes," Dean Moore said. "It is one more example of how climate change has transformed the moral landscape, and the emotional landscape. Do we have to find a way to live sustainably without dying of grief? As things we love are being destroyed?"

Eventually the sweet autumn rains came and doused the fire. In the new year the rivers and forests would awaken. New life would start in the burned-over, blackened landscapes of the Andrews. Soon, scientists would be out in the forest, repairing research plots, seeing what was burned, what they can learn. In a region where so much old-growth has been lost, said Matt Betts, the lead principal investigator at the Andrews, it was sad to think of having lost even more. He studies birds. How many would return to find no nest—their tree burned? This forest now had a new mission for the Andrews's scientists. It would be the premier place, with so much baseline data, to understand what happens when a forest burns. Even five-hundred-year old-growth. To me, this fire marked a turning point. To go to the Andrews now is to acknowledge both how much we have accomplished in understanding forests and to confront how much work we have to do to sustain and care for them.

First, the accomplishments. We understand forests today better than we ever have. Look at the change in our knowledge and understanding in one generation. We have moved, at least on some lands in the United States, from an era of cut and run and full-on exploitation to a more environmentally aware era, with an appreciation of forests as more than just places for producing wood. Today, in the best circumstances, where they are cut, forests also are managed not only for economic return but for preservation of their ecological function and capacity for renewal. Forests today, where they are appreciated and managed as they should be, are not just places for making money but also places we value for storing carbon and sheltering biodiversity and cultural uses—and places of intrinsic value we can't imagine living without. The dramatic change this encompasses, *just since the 1970s*, cannot be overstated, Jerry Franklin wrote to me in a letter, as we were taking stock at the end of this project of all we had learned and discussed.

"We have made tremendous progress," Franklin wrote. "Progress in our understanding of what a forest ecosystem is and how it works. That really did emerge from the science of the seventies, and the impact of that knowledge has been very powerful in both informing and implementing new policies." I, and we all, have reasons to be hopeful about our future, and the future of our forests, Franklin continued. "Who would have imagined in 1970 that we would essentially stop the harvest of old-growth forest on federal lands! I sure couldn't imagine it. Or that we would come up with a whole new philosophy about how to manage forests so as to sustain the full array of values. As a result of the knowledge that has been created, we are adopting and following very different strategies than previously. That is progress and it is hopeful."

Step back, he urged. Take a longer view. See how far we have come in our thinking and actions. "With my students I always tried to provide them with the perspective of time," he wrote. "In the early 1970s 'old-growth' was a swear word. Deep, dank, dark, decadent forests falling apart that need to be cut down and replaced with plantations. Now no one argues the merits of old-growth, although they might insist that they need to be cut in order to provide jobs . . . we have come an immense distance in the last sixty years in both knowledge and practice." He's right. While there is still so much more to do, we have seen in our own lifetimes that it is possible for science to inform policy, for people to learn and effect change. So what now?

THERE IS NO ONE conservation solution for a country like ours, with such dramatic differences in history, landscapes, and landownership. The reason we could save millions of acres of old-growth across three states in the Pacific Northwest is because it was under a single ownership—the federal government—which permitted a single solution over a vast area in an unprecedented effort, outside establishment of a national park, to safeguard biodiversity and old-growth. The complexities of charting a future in a place like Maine, where the forests are nearly entirely privately owned, are daunting. Or in Massachusetts, where there are some 293,000 small forest landowners managing some 1.6 million acres of forest. But solutions are emerging. In a provocative March 2024 publication *Beyond the Illusion of Preservation*, one way forward is presented by the report's seven authors, from a range of research and academic institutions. They call for a three-prong strategy: (1) Permanent protection of broad swaths of forestland, with a portion set aside for wildlands and the rest devoted to sustainable forestry. (2) Significantly reducing consumption of forest products. And (3) expanding ecological forestry, to sustainably grow and process wood products *locally*. The report is a prescription for New England forests. But its principles to me have broader application.

This is a vision that honors restraint on the consumption side—and rebuilding a local wood industry to build up the supply side, *from local sources*. Preserving wildlands to be havens of nature, as well as forestlands harvestable for local needs, to me is balanced, real, and honest—not fake conservation that looks the other way at cutting that just happens somewhere else beyond protected areas. Out of sight, out of mind. But still happening. This is forestry that reorients production toward higher-quality products, such as lumber, while sustaining ecological values—including on private, corporate-owned forests. This means corporate owners would in general have to cut less, and family forest owners would likely have to cut more, to meet the needs of the region. But it is a holistic and intentional approach to keep ecological, economic, and social values provided by forests in the picture, without sacrificing any to the other. This is not a new idea—it is a page from long-standing practice in traditional management systems. And it is at work on the ground today in, of all places, the nation's oldest outdoor club.

This is the Appalachian Mountain Club (AMC), in its Maine Woods Initiative, and the logging that the Londons and their crews are doing for the AMC, to regenerate the dense, even-age doghair stands left behind by industrial cutting. The AMC's goal is to generate revenue for its programs with higher-value forest products, including saw logs, while growing the forest into bigger trees and a more natural and healthy condition. The AMC cuts conservatively, with annual harvest that is about half the rate of the forest's annual growth. The wood is harvested by local crews and goes to local mills. What interested me about the AMC is its multiple values pursued on these lands—guaranteed public access for recreation, keeping the land on the local tax rolls, and investing in the future of the ecology and local economies of Maine. The AMC initiative is showing it is possible to do all of this— while managing nearly half its lands as wilderness areas, designated as a permanent ecological reserve. In those lands, natural succession in the forest will, over time, restore the complex old forest that once dominated the landscape.

Yes, the scale of all this is small—compared to the need. While 81 percent of New England is forested, only 3.3 percent has been permanently protected as wildlands, according to the sweeping 2023 report *Wildlands in New England: Past, Present and Future,* by a collaboration of more than one hundred partners including conservation organizations; municipal, state, and federal agencies; and several partners such as the Harvard Forest, the Highstead Foundation, and the Northeast Wilderness Trust. The report documented all of the wildlands of the New England region of every sort—the first such accounting of its kind. The authors proposed permanent protection for all existing farmland, and protection for at least 70 percent of the landscape as forest, with 10 percent of that set aside as wildlands. This is a vision that keeps people working in forests and farms to provide regionally produced food and fiber but also leaves some places to just unspool their long cadences of time, forever.

This is an ambitious vision, to be sure, but it is not unprecedented. New York's Adirondack Park was created in 1892 as New Yorkers watched their forests fall to the saw. At 6 million acres, today the park is the largest publicly protected area in the contiguous United States. That includes not only 2.6 million acres of wildlands forever set aside and owned by the state of New York, but another 3.4 million acres devoted to forestry, agriculture, and open space recreation. Much of the park's protected forests are not primary woodlands never cut—that's a rare thing on the East Coast. But what's there today in the preserved areas is conserved forever. It will be the old-growth of tomorrow.

These initiatives each are versions of a similar vision—*actual* conservation that protects forests permanently, leaving some in a wild condition, while carefully using the rest to create a flow of locally-sourced and cut and milled wood, sustaining jobs and local communities. Wouldn't that be straightforward and useful—to reduce our consumption, to responsibly and locally make what we use, and to keep wild as much of the rest as we can? We already celebrate local farmers, farms, and food production. Why not local woodlots and the people that cut and make the products we use? Where is the call for local wood and forestry equal to the appreciation for farmers and local food?

There's rethinking to do on the consumption side too. What if we insisted our local public buildings were made from local wood that was actually sustainably sourced, not just clear-cut. What if we took pride in not the big beams from century-old trees in grand and even second homes but thrifty, small homes made with wood recycled from teardowns and built with innovative products made from lower-value wood, such as wood fiber insulation? Can't we create tiny house envy, instead of lust for bigger, and even second homes? Can't we get more excited about a really big tree left in the forest for wildlife than we do about yet another, even bigger house for us? Whatever happened to the less-is-more movement of the 1970s? It has become virtually unallowed speech, in a consumption-oriented society that lusts for stuff, yet is disconnected from our own personal consumption. That we would even think about the wood products we use at all, or the people who make them, *and the forests they come from*, would be a good start. The Londons' crew was right to tease me about toilet paper. Forests are precious. We should treat them that way and honor the people who do this work well. Insist on forest practices that retain the biological and ecological health of the land and forests' capacity for renewal. Management practices that result not only in new homes for us but in retaining the homes of the animals. This is a forestry for all beings. A forestry that emphasizes leaving more, not taking more. A forestry for our future.

New beginnings are stirring. Their success is certain. The resilience of nature, if allowed to function, is 100 percent reliable. Witness the salmon, the herring, and the lamprey returning on both coasts to waters where they have not been in more than a century, the forests grown back where before there were clear-cuts. Forests that today are full of animals, life, and wonder. Much of the land originally cleared in southern New England today has regrown, in one of the greatest reforestations anywhere in the world. This was not by design but caused by people walking away from the plow as prime agricultural markets developed in the Midwest, with the opening of the Erie

Canal. With the return of the forests have come the animals. Today there are more people living closer to more wildlife in New England than in the time of Thoreau. There are fisher cats lounging in the second growth, and bobcats sallying along the tops of stone walls dividing what used to be pastures, today grown up in glorious oak, maple, white pine, beech, and birch.

In a beautiful piece published in 1995 in the *Atlantic Monthly*, environmentalist Bill McKibben in "An Explosion of Green" described the great rewilding of New England. "The proof that what is happening is significant lies in the recovery not only of the forests themselves, but of much of the life they always supported," he wrote. "Perhaps 40,000 black bears roam the East. In 1972, 37 wild turkeys were introduced into western Massachusetts. Today the population exceeds 10,000." Remember, he wrote that in 1995. *Today* the state of Massachusetts estimates wild turkey populations are between thirty thousand to thirty-five thousand, and the wild turkey is now the state game bird. Bears? They have become common throughout the Northeast, with an estimated thirty-five thousand black bears in the state of Maine alone—and nearly five thousand in Massachusetts. So many that people, even suburbanites, are learning to bring in the bird feeder at night. That's a lot of rewilding. Imagine what we could do if we worked even harder at it? That this is happening, in so many places, and in so many ways, and continuing in forests, streams, and rivers hammered in the first round of colonization and industrialization, is a definition of progress, for *our* time. I propose we also adopt an ethos of conservation for our time, based on reciprocity and respect in our relations with one another, and with nature. The rules of this ethos are simple: *Stop breaking things. Start fixing things. Stay together. Keep the big picture and the long game in focus. Share the cost and the work. And keep on going.*

Everywhere, at every scale, and do it joyfully. This is home keeping. This is community making. This is an ethos of conservation for today that includes and celebrates the role of people too, right at the center of nature, tending, caretaking, and using, yes, but with an eye always to

Penobscot elder and artist Butch Phillips sells his work at the Wabanaki Winter Market at the University of Maine in Orono in December 2022.

All My Relations, by Penobscot elder Butch Phillips, reflects the artist's teachings that all creation is one community—and that humans must live in a relationship of respect and reciprocity with nature.

the needs of next generation of people and all the nonhuman beings. It sounds simple but taking seriously an ethos of respect, interdependence, and reciprocity is powerful. I think of the words of Nuchatlaht elder and councillor Archie Little: everything is connected. I look often at a piece of artwork that Butch Phillips sold to me that day we sat together at the Wabanaki Winter Market, the art fair held at the University of Maine in Orono at the winter holiday time.

Artists had come from all over the region to sell everything from modest pieces to works of art acquired by museums. I pulled up a chair next to Phillips and we talked as he briskly sold his work. I bought a small and simple birch bark panel, framed with sticks. Carefully carved, barely more than scored, through a light layer of birch bark to reveal a second, darker layer below, Phillips had patiently rendered only by knife point a delicate portrait of Mount Katahdin, surrounded by a menagerie of tiny figures of woodland animals: moose, bear, deer, squirrel, fox, rabbit, owl, mink, beaver, of course trees, *and a person.* The title of the piece, handwritten on the back by Phillips, who also signed it, is *All My Relations.* And this, Phillips says, is the stance we must think about when we think now about surviving. Not just us, but *all* of us. All of the people and our nonhuman relations. For in taking care of the salmon rivers and salmon forests and all our relations, we take care of ourselves. There isn't really any choice.

Here is the other thing I learned from Phillips, age eighty-two and a member of the Penobscot Nation, which has survived in their same home territory more than twelve thousand years. Change in how we all are now living here together is both possible and necessary. And it won't happen on its own. "Until people accept there's a problem, and are willing to do something about it, and to accept we are damaging our future, until we accept that, there's not going to be any progress," Phillips told me. "But they need to understand, we are all connected to Mother Earth and any damage we do to Mother Earth, we do to ourselves. . . . The circle of life, the animals, the plants, the insects, the trees, everything is in this circle. And it is all connected."

ACKNOWLEDGMENTS

I AM SO GRATEFUL to so many people who made this book possible, and the doing of it so rich and fun. First, thanks to everyone at the University of Washington Press, especially its director, Nicole Mitchell, who caught the book's vision from our initial conversation and saw it to completion, with stalwart efforts from editorial director Larin McLaughlin and Beth Fuget, who managed fundraising for its publication. My agent, Elizabeth Wales, cheered me on. My colleagues at the *Seattle Times*, especially editors Ben Woodard and Matt Canham, made possible my absences from daily journalism to report and write this book and supported me in every way along a long journey.

The wonderful community at the Harvard Forest honored this project with a Bullard Fellowship in Forest Research to help support travel, research, and writing, much of it in their good company while I was eleven months in residence in Petersham, Massachusetts. Director Noel Michele Holbrook offered friendship and encouragement. Senior ecologists David Orwig and Neil Pederson generously took me with them on fieldwork amid old-growth trees in the Northeast and their insights fill this book. Laura Smith in the Tree Ring Lab tutored me in the processing and reading of tree cores and made the terrific suggestion of following one core from the tree in the field to the microscope in the lab. Senior ecologist Jonathan Thompson joined me in a small plane over the northern forest, just one of the numerous times he helped guide my appreciation of forests, their history, and future. Our weekly check-ins on walks in the forest helped me stay on track in a fast-developing project. David Foster, director at the Harvard Forest for thirty years from 1990 to 2020, encouraged me early on to explore the northern forest and introduced me to Mitch Lansky in Maine and Jamie Sayen in New Hampshire, authors of important books about the region's logging and paper industry. We three shared important conversations and walks that helped shape this book.

The Harvard Forest's campus and four thousand acres of woodlands, fields, trails, old stone walls, streams, and ponds, for more than a span of ten years and two fellowship residences, has come to feel like a second home. Its administrative staff have accommodated me and others in every imaginable and sometimes unimaginable need. For many years the Harvard Forest has been passionately curated by its Woods Crew, and their sage leader, my special mentor John Wisnewski, now retired to his own farm and family, knows and teaches from life many deep and wonderous things about forests, fields, and people far beyond what several PhDs would ever have gained for him.

In Petersham the Harvard Forest, far from its Cambridge mothership, is part of the equally special North Quabbin realm of Massachusetts. In Petersham and Athol and its environs are where living this book project was enriched with public libraries, a terrific YMCA, a rich historical and contemporary setting, and now many friendships. Journalist and author Allen Young and long-term Royalston civic stalwarts Carla and Phil Rabinowitz, to whom my husband introduced me a decade ago, have warmed all my intervening North Quabbin sojourns. Just as others of their circle, Gordon and Cynthia Donaldson in Lemoine, Maine, made me at home and gave me excellent counsel when the project required lengthy trips to Maine. John O'Keefe has been a dear friend and source of insight and inspiration since I first showed up at the Harvard Forest in 2013. Larry Buell generously taught me about the history of the North Quabbin.

Jerry Franklin, scientist, teacher, and policy advocate, has spent a remarkable career deeply invested in and contributing the larger scientific understandings this book relates. Many would say Jerry himself is the personification of an old-growth tree. My gratitude to him is huge. Jerry stuck with me on this project from our first conversations about an idea for a book to the very last steps—and all the changes, new directions, and rethinks in between. In the walks and talks he has shared with me both deep in the Cascade forests and in scalped clear-cuts, often joined by his companion, Edith Lindner, he taught me so much. Very special was the two weeks Jerry and Edith spent with me at the Harvard Forest, including a reunion with Jerry's friend and mine, the eminent international ecologist, author, and teacher Richard T. T. Forman of the Harvard University Graduate School of Design. I have been very lucky to engage them on this project, and I hope the product will merit what they have given it.

In the same category are scientists Fred Swanson and Julia Jones of Corvallis, Oregon. Their interest and enthusiasm buoyed this project even before I understood it would be a book. Not only for that and for all their own contributed insights, I am also grateful to them for introducing me to the special community where so much of their own work has centered at the H. J. Andrews Experimental Forest. There I was lucky to find director Mark Schulze, who in addition to sharing his knowledge and insight also understood how important it was for me to climb with him into the upper branches of the Discovery Tree. Mark Harmon, emeritus professor at Oregon State University, led me into the critical topic of forest decomposition and shared with me what he was going through as the Lookout Creek Fire descended on his life's work in July 2023. On the deeply personal sharing of that sorrowful catastrophe, thanks to Matt Betts and Brooke Penaluna, scientists at the Andrews, and author and philosopher Kathleen Moore for contributing with open hearts to a facet of this book no one saw coming.

I have so much appreciation for people who made my work in Maine so interesting and fruitful. That would certainly include Rory Saunders, fisheries biologist for NOAA, who offered research materials, contact information, good ideas, and encouragement—and time to take me on the water, for which there is no substitute. John Kocik, supervisory research fishery biologist, led me through the story of Penobscot River recovery, with much already achieved and more still to be accomplished. Justin Stevens of Maine Sea Grant was a wonderful guide to the recovering river seen during a spring fish survey. In everything connected with NOAA in Maine, I owe big thanks to Michael Milstein, spokesman for NOAA's West Coast Region, who helped open doors for me three thousand miles from home.

Other thanks are due to my own personal Maine guides, a very ancient and storied fraternity. Steve Tatko showed me the Golden Road and the Appalachian Mountain Club (AMC) vision of conservation and multiple use, even arranging a stay for me at the AMC's more than 140-year-old Little Lyford Lodge in Greenville, Maine. Molly and Alex London could not have been more generous in introducing me to their crew, and letting me see and experience for myself what logging looks like today in Maine. This is not easy work, and a largely untold story. I am grateful for all they taught me and for sharing their pride in their work and hopes for the future. Bill London showed me through his eyes forests he worked in for so many years and his pride in the bridges he is now building to help sea-run fish make it home. His excitement in seeing these fish in the forests of interior Maine—five dams from the sea—was crucial to this story.

It was Jessica Leahy, professor of forestry at the University of Maine School of Forest Resources, who introduced me to the Londons. She and her husband, Bob Seymour, emeritus professor of silviculture at the University of Maine, took me to the Penobscot Experimental Forest, to see a vision of what ecological forestry practices based on patterns of natural disturbance look like. Aaron Weiskittel at the University of Maine School of Forest Resources connected me to a wealth of people, research papers, and scientists who helped tell this story. Darren Ranco at the Wabanaki Center was welcoming and generous in sharing ideas and papers that contributed to my understanding. The staff at Brookfield Renewable Partners showed me Atlantic salmon up close at the fish lift at the Milford Dam, an unforgettable experience.

Other crucial contributions to the Maine chapters of this book were made by natural resource experts and others at the Penobscot Nation.

They were generous not only with expertise but by personally guiding me to forests, rivers, and streams in their trust lands and traditional territory. Dan Kusnierz, with thirty years of experience as the tribe's water resources program manager, explained as few else could have the legacy of pollution in the Penobscot River. Dan McCaw, fisheries program manager for the tribe, generously took me to the Penobscot River and its tributaries to learn the history of this river, its restoration, and its remaining needs. Charles Loring Jr., natural resources director for the tribe, and Ben Stevens, forestry director, explained the approach the tribe takes to forestry, and not only took me among the trees but also joined me up into the sky for an invaluable forest overview from a small chartered float plane.

Jason Mitchell, another of the tribe's longtime water quality experts, helped me understand the crucial link between tribal members' health and their ability to eat their First Foods, harvested again from their river. John Banks, the long-standing natural resources director for the tribe and a leader in the Penobscot restoration effort, shared important stories of the restoration, all underscoring the indispensable need for people to collaborate to make paths to better futures. Carol Dana, a Penobscot elder, became an important new friend and mentor, and wove a precious basket I now cherish. She welcomed me, shared meals together, and helped me think through what I was seeing and learning in her homeland. Penobscot elder Butch Phillips framed a central theme of this book by sharing what he and his tribe have seen, learned, and taught: reciprocity in caring for nature will be returned by nature taking care of us.

In British Columbia, hereditary chief Rande Cook of the Ma'amtagila First Nation invited me for a week of study and listening in his second annual Tree of Life gathering on Vancouver Island, an invaluable opportunity to spend time with members of his family from the Kwakwaka'wakw territories in the old-growth forests in their homeland. The Mother Tree Project team and their leader, scientist and author Suzanne Simard, welcomed me to watch them work as they study and struggle with the tenuous future of these forests. Teresa Ryan, a Gitlan/Tsimshian fisheries scientist, spoke unforgettably as we sat by the Tsitika River, and I hope to have done justice to her remarkable words and thoughts.

At the Mowachaht/Muchalaht Nation, elder and hereditary chief Jerry Jack Jr. spent a day with me explaining the Salmon Parks campaign and why it is so important. The family of elder Ray Williams graciously made space for a precious interview that barely preceded his walking on from this life. Archaeologist Jacob Earnshaw showed me culturally modified trees and generously shared his Vancouver Island and Nootka Island research. Anthropologist Chelsey Geralda Armstrong confronted false myths that obscure the true history of colonization of the First Nations of British Columbia. Roger Dunlap, formerly lands and natural resources manager for the Mowachaht/Muchalaht, and now working for Nuchatlaht, bushwhacked with me through the old-growth of Nootka Island in search of culturally modified trees. He shared the story of the Salmon Parks, and the Mowachaht/Muchalaht people's survival since the arrival of Captain Cook to the present day, an incredible saga of perseverance. All of that was richly reinforced and personally related by Nuchatlaht elder and councillor Archie Little.

Whenever my work takes me to new places and people, I always gratefully carry with me the friendship and guidance of my friend, Danita Washington, an esteemed elder of the Coast Salish Lummi Nation. My husband, Doug MacDonald, was on this book journey from beginning to end. He was a cheerful and observant companion on some of the reporting and stayed home to feed the cat and take care of our home when I was away. He read attentively every word of the emerging manuscript and offered helpful suggestions now seamlessly woven into the book. Years ago he placed on my desk, where it has stayed ever since, a well-worn copy of Marie Hall Ets's 1945 children's classic, *In the Forest* (which received a Caldecott Honor for its illustrations). It suitably anticipated the magic of this book's journey.

The Trees Are Speaking was made possible in part by the Tulalip
Tribes Charitable Fund, which provides the opportunity for a
sustainable and healthy community for all.

This book was also supported by a grant from the Hugh and
Jane Ferguson Foundation.

Additional support was provided by generous gifts from the
following individual donors:

Michael and Virginia Barry in honor of Todd Summerfelt
David S. Brown Jr.
Michael Burnap and Irene Tanabe
Debra Dahlen and Robert Fries
William Donnelly
Kelby Fletcher and Janet Boguch
Mimi Gardner Gates
Marcy Golde
Julia Jones and Fred Swanson
Martha Kongsgaard and Peter Goldman
Andrew Larson
Tim Ragen
Sonya Schneider and Stuart Nagae
Cynthia Sears
Claudia Vernia and Gail Gibson
Maggie Walker

Preface

The statistic on the amount of fossil fuel emissions absorbed by photosynthesis in the terrestrial ecosystem comes from Sophie Ruehr et al., "Evidence and Attribution of the Enhanced Land Carbon Sink," *Nature Reviews Earth & Environment* 4 (2023): 518–34, http://www.nature.com/articles/s43017-023-00456-3. The paper on the carbon density of temperate rain forests is by Heather Keith, Brendan G. Mackey, and David B. Lindenmayer, "Re-evaluation of Forest Biomass Carbon Stocks and Lessons from the World's Most Carbon-Dense Forests," *PNAS* 106, no. 28 (2009): 11635–40, http://www.pnas.org/doi/pdf/10.1073/pnas.0901970106.

Thomas E. Reimchen's work on tracing the nutrients of salmon into trees and wildlife in the salmon forests is "Diverse Ecological Pathways of Salmon Nutrients Through an Intact Marine-terrestrial Interface," Department of Biology, University of Victoria, https://doi.org/10.22621/cfn.v131i4.1965; and "Salmon Nutrients, Nitrogen Isotopes and Coastal Forests," *Ecoforestry* (Fall 2001), https://web.uvic.ca/~reimlab/reimchen_ecoforestry.pdf. There's much more on Reimchen's fascinating website, https://web.uvic.ca/~reimlab/.

John Reynold's work on seeing the influence of salmon in the plant communities of salmon streams and even the greenness of the trees is explored by Christopher J. Brown, Brett Parker, Morgan D. Hocking, and John D. Reynolds, "Salmon Abundance and Patterns of Forest Greenness As Measured by Satellite Imagery," *Science of the Total Environment* 725 (2020): 138448, http://www.sciencedirect.com/science

/article/abs/pii/S0048969720319616; and Morgan D. Hocking and John D. Reynolds, "Impacts of Salmon on Riparian Plant Diversity," *Science* 331 (2011), https://morganhocking.wordpress.com/wp-content /uploads/2013/10/hocking-and-reynolds-2011.pdf.

The quote from Thomas H. DeLuca and Michael Paul Nelson comes from their unpublished op-ed, "After the Fire, We Must Listen to the Trees," https://www.documentcloud.org/documents/24806093-oped _forests-burning_600-word-version_10-17-23-copy. They also inspired the title of this book.

1. Combing the High Winds

Readers wanting to experience the Cedar Flats Research Natural Area for themselves will find it easy to do. The one-mile loop trail is flat and easy, the experience spectacular. The forest trailhead is easily reached on a paved road. The nearest town with services is Cougar, Washington. For more information, see "Cedar Flats," Forest Service, USDA, https://research.fs.usda.gov/pnw/rnas/locations/cedar-flats.

For an in-depth history and look at the Andrews, see William G. Robbins, *A Place for Inquiry, a Place for Wonder: The Andrews Forest* (Corvallis: Oregon State University Press, 2020). For more on the old-growth ecosystems and the Andrews, see Jon R. Luoma, *The Hidden Forest: The Biography of an Ecosystem* (New York: Henry Holt, 1999). The H. J. Andrews Experimental Forest closed to public visits because of a 2023 wildfire. For current access information, see "Visit the Forest," https://andrewsforest.oregonstate.edu/about/visitor.

2. What Is an Old-Growth Forest?

The details of the characteristics of an old-growth forest come from the landmark paper on this topic by Jerry F. Franklin et al., "Ecological Characteristics of Old-growth Douglas-fir Forests," General Technical Report PNW-GTR-118 (Portland, OR: US Department of Agriculture, Forest Service, Pacific Northwest Research Station, 1981), https://research.fs.usda.gov/treesearch/5546. Also see Beverly Law et al., "Creating a Strategic Carbon Reserve Using Forests to Blunt Climate Change Effects," *Land* 11, no. 5 (2022), special issue, Forests in the Landscape Threats and Opportunities.

One of the earliest and still one of the best papers on the loss of carbon storage in forests when they are clear-cut is by Mark E. Harmon, William K. Ferrell, and Jerry F. Franklin, "Effects on Carbon Storage of Conversion of Old-Growth Forests to Young Forests," *Science* 247 (1990): 699–702, https://andrewsforest.oregonstate.edu/publications/1046. The zombie theory that "thrifty young stands" are where the most valuable carbon sequestration happens is still alive and well despite publication of this paper in 1990 and many subsequent papers.

Another excellent paper on the carbon storage capacity of the old forests of the Pacific Northwest—among the most carbon dense on the planet—is Beverly Law et al., "Land Use Strategies to Mitigate Climate Change in Carbon Dense Temperate Forests," *PNAS*, March 19, 2019, https://www.pnas.org/doi/10.1073/pnas.1720064115. A useful paper I have returned to over and over about the carbon sequestration sink in the world's forests is by Yude Pan et al., "A Large and Persistent Carbon Sink in the World's Forests," *Science* 333 (2011): 988–93, https://www.fs.usda.gov/research/treesearch/38624. For a sobering look at the decline of large old trees, see David Lindenmayer, William Laurance, and Jerry Franklin, "Global Decline in Large Old Trees," *Science* 338 (2012): 1305–56, https://doi.org.10.1126/science.1231070.

For photography of old-growth landscapes in the Pacific Northwest, and gorgeous portraits of loggers at work, enjoy the work of photographer David Paul Bayles, https://www.davidpaulbayles.com/. For a beautiful essay on old-growth forests, including photos by David Paul Bayles, see *A Cathedral Not Made by Hands* by Vincent Miller and published in *Commonweal Magazine* (December 2019), https://www.commonwealmagazine.org/cathedral-not-made-hands. For an intimate look at old-growth ecosystems in the Pacific Northwest accompanied by the extraordinary photography of Johsel Namkung, see Ruth Kirk with Jerry Franklin, *The Olympic Rain Forest: An Ecological Web* (Seattle: University of Washington Press, 1966).

I first visited the Cedar Flats Research Natural Area with photographer Steve Ringman and videographer Lauren Frohne for a story in the *Seattle Times Pacific NW Magazine.* Steve's photography and Lauren's video are not to be missed. See Lynda V. Mapes, "Meet the Eminent Scientist, Now 84, Who Vowed as a Boy to Protect Washington's Old-Growth Forests," *Seattle Times*, July 18, 2021, https://www.seattletimes .com/pacific-nw-magazine/reign-of-elders-washingtons-old-growth -forests-and-the-eminent-scientist-now-84-who-vowed-as-a-boy-to -protect-them/.

3. Drawing the Line

To learn more about the scientific takeaways from the Mount St. Helens eruption, I recommend Eric Wagner, *After the Blast: The Ecological Recovery of Mount St. Helens* (Seattle: University of Washington Press, 2020). The citation for the first paper on the hydrological effects of clear-cutting is Timothy D. Perry and Julia Jones, "Summer Streamflow Deficits from Regenerating Douglas-Fir Forest in the Pacific Northwest, USA," *Ecohydrology* 10, no. 2 (2017): 1–13, https:// doi.org/10.1002/eco.1790.

The second paper is by Catalina Segura, Kevin D. Bladon, Jeff A. Hatten, Julia A. Jones, V. Cody Hale, and George G. Ice, "Long-term Effects of Forest Harvesting on Summer Low Flow Deficits in the Coast Range of Oregon," *Journal of Hydrology* 585 (February 2020): 124749, https://doi.org/10.1016/j.jhydrol.2020.124749. The definitive history of the Northwest Forest Plan is the new book by K. Norman Johnson, Jerry F. Franklin, and Gordon H. Reeves, *The Making of the Northwest Forest Plan: The Wild Science of Saving Old Growth Ecosystems* (Corvallis: Oregon State University Press, 2023). For an in-depth exploration of forest ecosystems, see Jerry F. Franklin, K. Norman Johnson, and Debora L. Johnson, *Ecological Forest Management* (Long Grove, IL: Waveland Press, 2018). The hand-drawn illustrations by Robert Van Pelt are a treat.

4. Salmon Forests

To learn more about Suzanne Simard's research and her remarkable life, read her book *Finding the Mother Tree: Discovering the Wisdom of the Forest* (New York: Alfred A. Knopf, 2021). Her lab keeps an updated website about its work and findings at the Mother Tree Project, https://mothertreeproject.org/. For a clear-eyed view of the roots of the Pacific Northwest salmon crisis, see Jim Lichatowich, *Salmon Without Rivers: A History of the Pacific Salmon Crisis* (Washington, DC: Island Press, 1999). The report on the interconnected biodiversity of salmon and all the species they support is C. J. Cederholm et al., *Pacific Salmon and Wildlife—Ecological Contexts, Relationships, and Implications for Management*, Special Edition Technical Report (Olympia: Washington Department of Fish and Wildlife, 2000), https://wdfw .wa.gov/sites/default/files/publications/00063/wdfw00063.pdf.

5. Big Trees Matter

The paper on forest gardens is by Chelsey Geralda Armstrong, Jacob Earnshaw, and Alex C. McAlvay, "Coupled Archaeological and Ecological Analyses Reveal Ancient Cultivation and Land Use in Nuchatlaht (Nuu-chah-nulth) Territories, Pacific Northwest," *Journal of Archaeological Science* 143 (2022): 05611, https://doi.org/10.1016/j .jas.2022.105611. J. K. Earnshaw's article on culturally modified trees is "Cultural Forests in Cross Section: Clear-Cuts Reveal 1,100 Years of Bark Harvesting on Vancouver Island, British Columbia," *American Antiquity* 84, no. 3 (2019): 516–30, https://doi.org/10.1017 /aaq.2019.29.

Maxine Berg's article is "Sea Otters and Iron: A Global Micro-history of Value and Exchange at Nootka Sound, 1774–1792," *Past & Present* 242, Supplement 14 (November 2019): 50–82, https://doi.org/10.1093/pastj/gtz038. I first reported on the Salmon Parks campaign in the *Seattle Times* in February 2023 with my colleagues photographer Erika Schultz and videographer Lauren Frohne. For their incredible photos and video, see "Salmon Parks: Inside a Movement to Conserve Pacific Northwest Old Growth," https://projects.seattletimes.com/2023/local/salmon-parks-movement-to-conserve-Pacific-Northwest-old-growth/.

The quote from Cook's journal is from "Whose Land Is It? Rethinking Sovereignty in British Columbia" by John Price and Nick Claxton in *BC Studies: The British Columbian Quarterly*, no. 204 (Winter 2019/20), https://doi.org/10.14288/bcs.v0i204.191508. The citation for the paper on infrequent fire history on Vancouver Island is by Daniel G. Gavin, Linda B. Brubaker, and Kenneth P. Lertzman, "Holocene Fire History of a Coastal Temperate Rain Forest Based on Soil Charcoal Radiocarbon Dates," *Ecology* 84, no. 1 (2003): 186–201, http://www.jstor.org/stable/3108008.

Readers will enjoy this piece on culturally modified trees: Lynda V. Mapes, "The Pacific Northwest Trees Shaped by Generations of People," *Seattle Times*, November 12, 2023, http://www.seattletimes.com/seattle-news/environment/the-pacific-northwest-trees-shaped-by-generations-of-people/. The video and photos by Lauren Frohne and Erika Schultz are beautiful, and the illustrations by Fiona Martin fantastic. The population estimates at Yuquot before colonization and history of displacement and forced removal come from "Yuquot: Where the Wind Blows from All Directions," an unpublished presentation by the Mowachaht/Muchalaht about their history, made in March 2022.

6. Mossy and Moosey

For the deep read, you still can't beat Henry David Thoreau, *The Maine Woods* (Boston: Ticknor and Fields, 1864). For a beautiful paper that tracks Thoreau's writings about the deforestation of Maine, read Geoffrey Paul Carpenter, "Deforestation in Nineteenth-Century Maine: The Record of Henry David Thoreau," *Maine History* 38, no. 1 (1998): 2–35, https://digitalcommons.library.umaine.edu/mainehistory journal/vol38/iss1.

I relied on the wonderful book by Richard G. Wood, *A History of Lumbering in Maine 1820–1861* (Orono: University of Maine Press, 1971), for details on what early logging was like in these woods. I relied on David Dobbs and Richard Ober, *The Northern Forest* (White River Junction, VT: Chelsea Green Publishing, 1995), for the history of the forest products industry in Maine. For a more contemporary account on the Maine woods, see Andrew M. Barton, with Alan S. White and Charles V. Cogbill, *The Changing Nature of the Maine Woods* (Durham: University of New Hampshire Press, 2012).

For the history of papermaking, I relied on Lloyd C. Irland's "Papermaking in Maine: Economic Trends from 1894 to 2000," *Maine History* 45, no. 1 (2009): 53–74, https://digitalcommons.library.umaine .edu/mainehistoryjournal/vol45/iss1/6. For a moving account of the importance of paper mill jobs in New England communities, read the one-of-a-kind book by Jamie Sayen, *You Had a Job for Life: Story of a Company Town* (Hanover, NH: University Press of New England, 2018), about the Groveton mills in New Hampshire and the aftermath of their shutdown. Also see Sayen's new book, *Children of the Northern Forest: Wild New England's History from Glaciers to Global Warming* (New Haven, CT: Yale University Press, 2023).

For another view, read Kerri Arsenault's *Mill Town: Reckoning with What Remains* (New York: St. Martin's Press, 2020), a moving primer by an author who grew up in a paper mill town family in Mexico, Maine, on how and why pollution is tolerated in one-industry towns, even when it causes cancer. For history of the Maine forests, I relied on Lloyd C. Irland, *The Northeast's Changing Forest* (Petersham, MA: Distributed by Harvard University Press for Harvard Forest, 1999). Another excellent paper on Maine forest history is Irland's "From Wilderness to Timberland to Vacationland to Ecosystem: Maine's Forests, 1820–2020," *Maine Policy Review* 29, no. 2 (2020): 45–56, https:// digitalcommons.library.umaine.edu/mpr/vol29/iss2/7/.

For a deep dive into the labor history of Maine's paper mill industry, read Michael G. Hillard, *Shredding Paper: The Rise and Fall of Maine's Mighty Paper Industry* (Ithaca, NY: Cornell University Press, 2020). For an overview of the financialization of Maine's timberlands and its impact, see A. Gunnoe and P. K. Gellert, "Financialization, Shareholder Value, and the Transformation of Timberland Ownership in the US," *Critical Sociology* 37, no. 3 (2011): 265–84, https:// doi.org/10.1177/0896920510378764. Also see A. Gunnoe, C. Bailey, and L. Ameyaw, "Millions of Acres, Billions of Trees: Socioecological Impacts of Shifting Timberland Ownership," *Rural Sociology* 83 (2018): 799–822, https://doi.org/10.1111/ruso.12210.

The paper about the degradation of New England forests is by John S. Gunn, Mark J. Ducey, and Ethan Belair, "Evaluating Degradation in a North American Temperate Forest," *Forest Ecology and Management* 432 (2019): 415–26, https://doi.org/10.1016/j .foreco.2018.09.046. For the definitive history of clear-cutting in Maine and the effect of industrial forestry on the people and land of the North Maine Woods, see Mitch Lansky, *Beyond the Beauty Strip: Saving What's Left of Our Forests* (Gardiner, ME: Tilbury House, 1992).

For more excellent reporting and writing on Maine forest history and the promise of low-impact forestry, see *Low-Impact Forestry: Forestry as if the Future Mattered*, edited by Mitch Lansky (Hallowell: Maine Environmental Policy Institute, 2002), http://planetmaine .net/meepi/lif/LIF%20Book.pdf. Also collected at this link are more of Lansky's articles and essays, including *Beyond the Beauty Strip: A 20th Year Retrospective*, a wonderful compilation of the history and state of the Maine woods and Lansky's home ground at Reed Plantation. Another excellent paper on the change in these woods is by Jonathan R. Thompson et al., "Four Centuries of Change in Northeastern United States Forests," *PLOS One* 8, no. 9 (September 2013), https://journals.plos.org/plosone/article?id=10.1371/journal.pone.0072540.

7. Stinkin' Lincoln

The profile of the paper industry in the state of Maine is from the "Pulp and Paper Mill Products Market Profile for 2002" by the Office of Business Development Domestic Trade Program in the Maine Department of Economic and Community Development, https://www.maine.gov/decd/sites/maine.gov.decd/files/inline-files/Market%20Profile%20%20-%20Pulp%20and%20Paper%20Products%20-%20State%20of%20Maine%20DECD.pdf.

The details of the effects of early pollution and alteration of the streams in the Penobscot watershed is from the report *Aroostook River Salmon Restoration and Fisheries Management,* by Kendall Warner, regional fishery biologist, Maine Department of Inland Fisheries and Game, 1956, https://www.nativefishlab.net/library/textpdf/14889.pdf. The report on cancer rates is from the Penobscot Nation, Department of Natural Resources, page 5, https://www.cclr.org/wp-content/uploads/2022/01/Penobscot-21-A-036_Redacted.pdf.

The history of the Wabanaki people and use of their First Foods and risk of pollution from the river now in their traditional foods is from the excellent and comprehensive report *Wabanaki Traditional Cultural Lifeway Exposure Scenario* (July 2009), https://www.epa.gov/sites/default/files/2015-08/documents/ditca.pdf, prepared for the US Environmental Protection Agency in collaboration with Maine tribes by Dr. Barbara Harper, emeritus faculty at the College of Public Health and Human Sciences at Oregon State University, and professor Darren Ranco, then at Dartmouth, now at the University of Maine where he is the director of the Wabanaki Center. Dr. Harper passed away in 2021. The 2021 Agency for Toxic Substances and Disease Registry report on the Penobscot River pollutants and risks to the tribe from eating its First Foods is at "ATSDR 2021 Annual Report," US Department of Health and Human Services, https://www.atsdr .cdc.gov/2021-annual-report/listening-responding-taking-action /penobscot-river.html.

For more on the early history of the log drives and Big Sabattus Mitchell, see Bill Caldwell, *Rivers of Fortune: Where Maine Tides and Money Flowed* (Portland, ME: G. Gannett Pub. Co., 1983). The citation for contamination of wildlife from pollution in the Penobscot is from Lisa Jo Melnyk et al., "Risks from Mercury in Anadromous Fish Collected from Penobscot River, Maine," *Science of the Total Environment* 781 (2021): 14669, https://doi.org/10.1016/j.scitotenv.2021.146691.

For more on the pollution of anadromous fish and health risks to the Penobscot people, see "Health Consultation: Review of Anadromous Fish in the Penobscot River," May 19, 2021, Agency for Toxic Substances and Disease Registry, US Department of Health and Human Services, https://www.atsdr.cdc.gov/hac/pha/PenobscotRiver /Penobscot_Indian_Nation_HC-508.pdf. For findings by the EPA, see "The Penobscot River and Environmental Contaminants: Assessment of Tribal Exposure Through Sustenance Lifeways," August 2015, Final RARE Report, EPA Office of Research and Development, https:// www.epa.gov/sites/default/files/2015-12/documents/final-rare-report -august-2015.pdf.

8. New Beginnings

Charles D. Canham's report on the problems with offsets can be found at "Rethinking Forest Carbon Offsets," June 15, 2021, https://www.cary institute.org/news-insights/feature/rethinking-forest-carbon-offsets. For more on the abrupt change to strip cutting in Maine, see K. R. Legaard, S. A. Sader, and E. M. Simons-Legaard, "Evaluating the Impact of Abrupt Changes in Forest Policy and Management Practices on Landscape Dynamics: Analysis of a Landsat Image Time Series in the Atlantic Northern Forest," *PLOS One* 10, no. 6 (2015): e0130428, https://doi.org/10.1371/journal.pone.0130428.

Also see "Long-term Outcomes and Tradeoffs of Forest Policy and Management Practices on the Broad-Scale Sustainability of Forest Resources: Wood Supply, Carbon, and Wildlife Habitat," principal investigator Erin Simons-Legaard, School of Forest Resources, University of Maine, Orono; coprincipal investigators Kasey Legaard, Jeremy Wilson, and Steve Sader; and collaborators Andrew Lister and Brian Sturtevant, https://nsrcforest.org/sites/default/files/uploads /simons-legaard10full.pdf.

For more on the importance of logging and trucking in Maine, see Megan Bailey and Andrew Crawley, "The Economic Contribution of Logging and Trucking in Maine," February 2023, EDA Center at the University of Maine, https://plcloggers.org/wp-content /uploads/2023/02/Logging-and-Trucking-Impact-02.01.23-1.pdf. The citation for the paper on easements is J. R. Thompson et al., "Do Working Forest Easements Work for Conservation," *Environmental Research Letters* 19, no. 11 (October 2024), https://iopscience.iop.org/ article/10.1088/1748-9326/ad7ed9.

9. *Aboard the* Silver Smolt

For a beautiful explanation of the intertwined suite of life in the Penobscot, see R. Saunders, M. A. Hachey, and C. W. Fay, "Maine's Diadromous Fish Community: Past, Present, and Implications for Atlantic Salmon Recovery," *Fisheries* 31 (2006): 537–47, https://doi.org /10.1577/1548-8446(2006)31[537:MDFC]2.0.CO;2. To understand just how important sea lamprey are as ecosystem engineers in the Penobscot, see R. S. Hogg, S. M. Coghlan Jr., J. Zydlewski, and K. S. Simon, "Anadromous Sea Lampreys (*Petromyzon marinus*) Are Ecosystem Engineers in a Spawning Tributary," *Freshwater Biology* 59 (2014): 1294–1307, https://doi.org/10.1111/fwb.12349.

The best account on the Penobscot River restoration project is by Peter Taylor, *From the Mountains to the Sea* (Yarmouth, ME: Islandport

Press, 2020). Taylor extensively interviewed the participants in this effort. For fish counts in the Penobscot River at the Milford Dam, the first dam fish cross, currently and over time, see "Trap Count Statistics," Maine Department of Marine Resources, https://www.maine.gov/dmr/fisheries/sea-run-fisheries/programs-and-projects/trap-count-statistics.

A helpful paper for understanding the Penobscot River restoration in terms of its success in fish recovery—and how much is yet to be done—is by Tara R. Trinko Lake, Kyle R. Ravana, and Rory Saunders, "Evaluating Changes in Diadromous Species Distributions and Habitat Accessibility Following the Penobscot River Restoration Project," *Marine and Coastal Fisheries* 4, no. 1 (2012): 284–93, https://doi.org/10.1080/19425120.2012.675971. There has been a lot of exaggeration even found in federal agency publications on the amount of habitat opened so far. There is a difference between fully open, unobstructed habitat and improved access, which this paper explains.

Another good assessment on the results of dam removal so far is in K. A. Whittum, J. D. Zydlewski, S. M. Coghlan Jr., D. B. Hayes, J. Watson, and I. Kiraly, "Fish Assemblages in the Penobscot River: A Decade after Dam Removal," *Marine and Coastal Fisheries* 15 (2023): e10227, https://doi.org/10.1002/mcf2.10227. For more on the estuary survey tracking biomass in the river, see Christine A. Lipsky, Rory Saunders, Justin R. Stevens, Michael O'Malley, and Paul Music, "Developing Sampling Strategies to Assess the Penobscot River Estuary (2010–2013)," 2019 Northeast Fisheries Science Center reference document, 19–02, https://doi.org/10.25923/xsj4-gz69.

10. The Trees Are Speaking

The paper on old-growth on Wachusett Mountain is by David Orwig, Charles Cogbill, David Foster, and John O' Keefe, "Variations in Old-Growth Structure and Definitions: Forest Dynamics on Wachusett Mountain, Massachusetts," *Ecological Applications* 11 (2001):

437–52, https://doi.org.10.1890/1051-0761(2001)011[0437:VIOGSA]
2.0.CO;2.

The final report on the surveys of old-growth on Wachusett
Mountain includes tables of all the species, with ages and sizes of
the trees. A feast for the imagination. See D. R. Foster, D. A. Orwig,
and J. F. O'Keefe, "Old-Growth Forest Monitoring on Wachusett
Mountain," September 22, 1997, https://www.documentcloud.org
/documents/24688937-ogreport-1997final. For an intriguing read
that challenges the oft-stated claim that chestnut was the dominant
tree in the eastern forests of the United States, see Edward K. Faison
and David R. Foster, "Did American Chestnut Really Dominate
the Eastern Forest?," October 2014, https://arboretum.harvard.edu
/stories/did-american-chestnut-really-dominate-the-eastern-forest/.
Beautifully written, and the historical photos of chestnut forests are
spectacular. Neil Pederson's guide to eastern old-growth is in *Natural
Areas Journal* 30, no. 4 (2010), https://doi.org/10.3375/043.030.0405.

For an excellent overview of four centuries of change in northeastern US forests, see J. R. Thompson, D. N. Carpenter, C. V. Cogbill, and
D. R. Foster, "Four Centuries of Change in Northeastern United States
Forests," *PLOS One* 8, no. 9 (2013): e72540, https://doi.org/10.1371
/journal.pone.0072540. The report on old-growth on Wachusett that
found beech bark disease prevalent is David Orwig, Charles Cogbill,
David Foster, and John O'Keefe, "Variations in Old-Growth Structure
and Definitions: Forest Dynamics on Wachusett Mountain, Massachusetts," *Ecological Applications* 11 (2001): 437–52, https://doi.org/10
.1890/1051-0761(2001)011[0437:VIOGSA]2.0.CO;2.

History on the chestnut blight is from Jesse D. Diller, "Chestnut
Blight," Forest Pest Leaflet 94, March 1965, US Department of Agriculture, https://www.fs.usda.gov/Internet/FSE_DOCUMENTS
/fsbdev2_043617.pdf. On the history of Dutch elm disease, see the
Morton Arboretum, https://mortonarb.org/plant-and-protect/tree
-plant-care/plant-care-resources/dutch-elm-disease/#overview;
and David Karnosky, "Dutch Elm Disease: A Review of the History,

Environmental Implications, Control, and Research Needs," *Environmental Conservation* 6, no. 4 (1979): 311–22, https://www.jstor.org/stable/44517039?seq=1. The connection between climate change and the spread of disease is from A. S. Weed, M. P. Ayres, and J. A. Hicke, "Consequences of Climate Change for Biotic Disturbances in North American Forests," *Ecological Monographs* 83 (2013): 441–70, https://doi.org/10.1890/13-0160.1.

The paper on live plant imports as a pathway for disease is by Andrew M. Liebhold et al., "Live Plant Imports: The Major Pathway for Forest Insect and Pathogen Invasions of the US," *Frontiers in Ecology and the Environment* 10 (2012): 135–43, https://doi.org/10.1890/110198. The connection between wood packing material and the spread of disease and the policy implications is well explained by Gary M. Lovett et al., "Nonnative Forest Insects and Pathogens in the United States: Impacts and Policy Options," *Ecological Applications* 26 (2016), https://www.caryinstitute.org/sites/default/files/downloads/tree_smart_trade_ecol_app_journal_impacts_policy.pdf.

The paper on the global importance of big trees is J. A. Lutz, T. J. Furniss, D. J. Johnson et al., "Global Importance of Large-diameter Trees," *Global Ecology and Biogeography* 27 (2018): 849–64, https://doi.org/10.1111/geb.12747. See more on Oregon State University scientist Beverly Law's argument that we should create a strategic carbon reserve in our nation's forests here: "Keeping Trees in the Ground Where They Are Already Growing Is an Effective Low-Tech Way to Slow Climate Change," *The Conversation*, February 23, 2021, https://theconversation.com/keeping-trees-in-the-ground-where-they-are-already-growing-is-an-effective-low-tech-way-to-slow-climate-change-154618. Her paper on this topic is Law et al., "Strategic Forest Reserves Can Protect Biodiversity in the Western United States and Mitigate Climate Change," *Communications Earth & Environment* 2 (2021), article 254, https://doi.org/10.1038/s43247-021-00326-0.

The paper on pro forestation is by William R. Moomaw, Susan A. Masino, and Edward K. Faison, "Intact Forests in the United States: Pro-forestation Mitigates Climate Change and Serves the Greatest Good," *Frontiers in Forests and Global Change* 2 (2019), https://www.frontiersin .org/articles/10.3389/ffgc.2019.00027/full. The paper that found one in four tree deaths was due to insects or disease in the study area in the Blue Ridge Mountains is by K. J. Anderson-Teixeira, V. Herrmann, W. B. Cass, et al., "Long-Term Impacts of Invasive Insects and Pathogens on Composition, Biomass, and Diversity of Forests in Virginia's Blue Ridge Mountains," *Ecosystems* 24 (2021): 89–105, https://doi .org/10.1007/s10021-020-00503-w.

The information on Wabanaki basketmaking comes from Jennifer Neptune and Lisa Neuman, "Basketry of the Wabanaki Indians," in *Encyclopaedia of the History of Science, Technology, and Medicine in Non-Western Culture*, https://umaine.edu/nativeamericanprograms/wp-content /uploads/sites/320/2017/07/Neuman-Basketry-of-the-Wabanaki -Indians.pdf. For more on black ash, see Kara K. L. Costanza et al., "The Precarious State of a Cultural Keystone Species: Tribal and Biological Assessments of the Role and Future of Black Ash," *Journal of Forestry* 115, no. 5 (September 2017): 435–46, https://doi.org/10.5849/ jof.2016-034R1.

For more on emerald ash borer, see Alex Smith, "Emerald Ash Borer at Hopewell Culture National Historical Park," National Park Service, https://www.nps.gov/articles/000/emerald-ash-borer -at-hopewell-culture-national-historical-park.htm. On the incredibly destructive effects of emerald ash borer, see "Emerald Ash Borer," Animal and Plant Health Inspection Service, USDA, June 7, 2024, https://www.aphis.usda.gov/aphis/ourfocus/planthealth/plant-pest -and-disease-programs/pests-and-diseases/emerald-ash-borer. For a good look at what happens to a forest when a pest wipes out a foundational tree species, see B. S. Case et al., "When a Foundation Crumbles: Forecasting Forest Dynamics Following the Decline of the Foundation Species *Tsuga canadensis*," *Ecosphere* 8, no. 7 (2017): e01893, https://doi.org.10.1002/ecs2.1893.

For more on the demise of chestnut, see Donald Edward Davis,

The American Chestnut: An Environmental History (Athens: University of Georgia Press, 2021). For an elegiac account of eastern hemlock and its demise, see David Foster, ed., *Hemlock: A Forest Giant on the Edge* (New Haven, CT: Yale University Press, 2014), with beautiful photography and a lovely design. Written by scientists and associates of the Harvard Forest, this book is a deep account, both scientifically rich and movingly written by scientists who deeply know and love this tree and these woods.

Portions of this chapter were first published in "What We Lose When We Lose Old-Growth Forests," *Atmos* 8 (2023), https://atmos .earth/what-we-lose-when-we-lose-old-growth-forests/. This new and superb long-form magazine with no advertising features environmental topics, with a global perspective on climate and culture, presented with extraordinary design and photography.

11. Home Keeping

For an elegant think piece on the spiritual values of old-growth, read Kathleen Dean Moore, "In the Shadow of the Cedars: The Spiritual Values of Old-Growth Forests," *Conservation Biology* 21, no. 4 (2007): 1120–23, https://www.jstor.org/stable/4620924. Oregon State University forest decomposition expert Mark Harmon and his collaborators remind us all that even when a forest burns in a megafire, it does not go up in smoke. Most of it is still standing. See Mark Harmon, Chad Hanson, and Dominick DellaSala, "Combustion of Aboveground Wood from Live Trees in Megafires, CA, USA," *Forests* 13, no. 391, https://doi.org.10.3390/f13030391.

The website of the Andrews has an extensive chronicle of the Lookout Fire, including photos, essays, and before-the-fire posts; richly detailed looks at what the forest and its plant and animal community was like before; and what scientists will be looking for now and in the future. See "Andrews Forest Program," H. J. Andrews Experimental Forest, https://andrewsforest.oregonstate.edu/. For a provocative and informative look at forests after the 2020 Holiday Farm Fire—one of the largest in Oregon history—see the art/science collaboration of David Paul Bayles and Fred Swanson at "Following Fire," https://www.followingfire.com/.

For fresh thinking on the future of wildlands and woodlands in New England, see Caitlin Littlefield et al., "Beyond the 'Illusion of Preservation': Taking Regional Responsibility by Protecting Forests, Reducing Consumption, and Expanding Ecological Forestry in New England," March 2024, https://masswoods.org/sites/default/files/Beyond-the-Illusion-of-Preservation-web.pdf. The "Wildlands and Woodlands" report and its conservation vision for New England is at Harvard Forest, https://harvardforest.fas.harvard.edu/other-tags/wildlands-woodlands. Bill McKibben's article "An Explosion of Green," *The Atlantic*, April 1995, is at https://www.theatlantic.com/magazine/archive/1995/04/an-explosion-of-green/305864/. See Erle C. Ellis et al., "People Have Shaped Most of Terrestrial Nature for at Least 12,000 Years," *PNAS*, April 19, 2021, https://www.pnas.org/doi/abs/10.1073/pnas.2023483118.

For all things northern forest, see the online scanned library of all editions of *The Northern Forest Forum*, a brilliant publication by local writers and artists of New England published from 1992 through 2002. Available on the Harvard Forest website at https://harvardforest.fas.harvard.edu/northern-forest-forum-1992-2002 and worth reading even more today, as each edition raises still unresolved issues about this landscape. The cartoons, polls, and artwork alone are a must-see. So are the quizzes. This journal underlays much of the intellectual history of the movement to conserve wildlands in Maine for the good of people, communities, and biodiversity. This conservation vision is moving forward today beyond Maine to New England, with momentum from the group Wildlands, Woodlands, Farmlands & Communities, https://wildlandsandwoodlands.org/.

INDEX

Page numbers in *italics* refer to illustrations.